Advanced Natural Gas Engineering

Advanced Natural Gas Engineering

Editor

Manoj Karkare

scitus
academics

Advanced Natural Gas Engineering

Edited by **Manoj Karkare**

Printed in 2017

ISBN: 978-1-68117-328-3

Library of Congress Control Number: 2015936524

SCITUS Academics LLC,
616, Corporate Way, Suite 2, 4766,
Valley Cottage, NY 10989

www.scitusacademics.com

Notice

Contents

Preface

Natural-gas processing is a complex industrial process designed to clean raw natural gas by separating impurities and various non-methane hydrocarbons and fluids to produce what is known as pipeline quality dry natural gas. Natural-gas processing begins at the well head. The composition of the raw natural gas extracted from producing wells depends on the type, depth, and location of the underground deposit and the geology of the area. Oil and natural gas are often found together in the same reservoir. The natural gas produced from oil wells is generally classified as associated-dissolved, meaning that the natural gas is associated with or dissolved in crude oil. Natural gas production absent any association with crude oil is classified as "non-associated." In 2009, 89 percent of U.S. wellhead production of natural gas was non-associated.

Editor

Chapter 1

Performance Evaluation of Refrigeration Units in Natural Gas Liquid Extraction Plant

Awajiogak Anthony Ujile and Dirina Amesi

Department of Chemical/Petrochemical Engineering, Faculty of Engineering, Rivers State University of Science and Technology, PMB 5080, Port Harcourt 50000, Nigeria

ABSTRACT

This paper has applied thermodynamics principles to evaluate the reliability of 390 m^3/hr natural gas processing plant. The thermodynamics equations were utilized in the evaluation, characterization, and numerical simulation of key process parameters in natural gas liquid extraction plant. The results obtained show the comparison of the coefficient of performance, compression ratio, isentropic work, actual work, electrical power requirements, cooling water consumption in intercoolers, compressor power output, compressor capacity, and isentropic, volumetric, and mechanical efficiency of the two-stage refrigeration unit with a flash gas economizer and these were compared with the designed specifications. The second law of thermodynamics was applied in analyzing the refrigeration unit and the result shows that exergetic losses or lost work due to irreversibility falls within operating limit that is less than 1.0%. Similarly, the performance of expansion turbine (expander) parameters was monitored and the results indicate a considerable decrease in turbine efficiencies as the inlet gas pressure increases resulting in an increased power output of the turbine leading to a higher liquefaction rate.

INTRODUCTION

The production and availability of natural gas liquid depend largely on the supply of raw natural gas from production wellhead and the operating conditions of the process unit that make up the extraction plant. Most gas processing plants are faced with problems ranging from inadequate supply, poor facility performance, and human factors. These problems can lead to low productivity of natural gas liquids and reduction in gas quality which could result in shutting down of the plant. Poor facilities such as inadequate electricity supply and processed water supply used in process equipment also lead to intermittent operations and malfunction of process equipment such as pumps, compressors, and valves if they are not adequately checked.

Human factors may also result from the inability of gas plant operators to monitor thermodynamics parameters such as the pressure, flow rates, and temperature on process equipment which could result in loss of data in control room and unintended activation or deactivation of process devices and reduce the plant efficiency. In spite of the fact that these thermodynamics parameters are monitored daily in a gas plant, there are problems of low gas feed inlet pressure and insufficient gas flow rate. These have resulted in low volume of natural gas liquids produced, the extracted natural gas liquids not attaining the expected cryogenic temperature requirement, and variation in gas quality discharging from the outlet of natural gas liquid extraction unit bottom product.

If the available inlet gas pressure is low, it can result in compressor system suction pressure falling below atmospheric pressure. This can also lead to air leakages into the compressor system, contributing to pulsation, corrosion, and low heating value of natural gas. In order to solve these problems, performance evaluation of the process units in natural gas liquids extraction using thermodynamics principles is necessary to ensure that these problems are minimized.

It is seen from literature survey that several papers have been published focusing on thermodynamic analysis of a gas turbine power plant.

Rahman et al. [1] and Taniguchi and Miyamae [2] carried out the study on the effects of ambient temperature and ambient pressure as well as temperature of exhaust gases on performance of gas turbine. Khaliq and Kaushik [3] studied the efficiency of a gas turbine cogeneration system with heat recovery steam generator.

Keith and Kenneth [4] developed a new method of applying overall plant material balance equation to determine the performance of a natural gas processing plant. Jibril et al. [5] studied the simulation of expansion turbine (expander) for recovery of natural gas liquids from natural gas stream gas using HYPRO TECH's HYSYS process simulation software. Donnelly et al. [6] carried out researches on process simulation and optimization of cryogenic operations using multistream brazed aluminum heat exchangers. Ganapathy et al. [7] studied the energy analysis of operating lignite fired thermal power plant. Design methodology for parametric study and thermodynamic performance evaluation of natural gas process plant has been developed. Refrigeration unit and expansion turbine unit which are the major components of the plant were evaluated. The expansion of natural gas through an expansion turbine with efficiencies shows a very good performance Ujile and Alawa [8]. The comparison between the energy losses and the energy losses of the components unit were investigated.

This paper examined the reliability of a gas plant which is the ability of a plant to maintain a stable efficiency with respect to time using thermodynamics equations.

METHODOLOGY/SYSTEM DESCRIPTIONS

Propane Refrigeration Unit

The simulation of process units in propane refrigeration cycle involves the applications of thermodynamics principles to the following.

- Calculate the amount of heat added to or removed from process streams.
- Estimate the power requirements for process equipment such as pumps, compressors, and turbine.
- Evaluate the performance of a flash separator at various temperatures and pressures.
- Determine the bubble and dew point temperature associated with distillation and bottom products.

A systematic diagram of propane refrigeration cycle is shown in Figure 1. The refrigerant passes through the scrubber into the first stage compressor at 1.9 bar as saturated vapor where it mixes up with the vapor from the economizer unit. It is compressed to the inlet of the second stage compressor at 2.5 bar and discharges at 5.0 bar into the condenser as superheated vapor. It leaves the condenser at 4.7 bar and enters the first expansion valve and from the expansion valve into an economizer drum where it flashes at an intermediate pressure of 3.3 bar. The vapour phase with the economizer at the top mixes up the vapour from the first stage compressor, while the liquid discharges at the bottom and enters the second expansion valve into the chiller at a pressure of 2.5 bar. The process gas (natural gas) flows inside the tubes connected to the chiller at 41 °C and countercurrently exchanges heat and gives up its energy with the liquid refrigerant surrounding the tubes. The refrigerant boils up and leaves the chiller space as saturated vapor to be recompressed again.

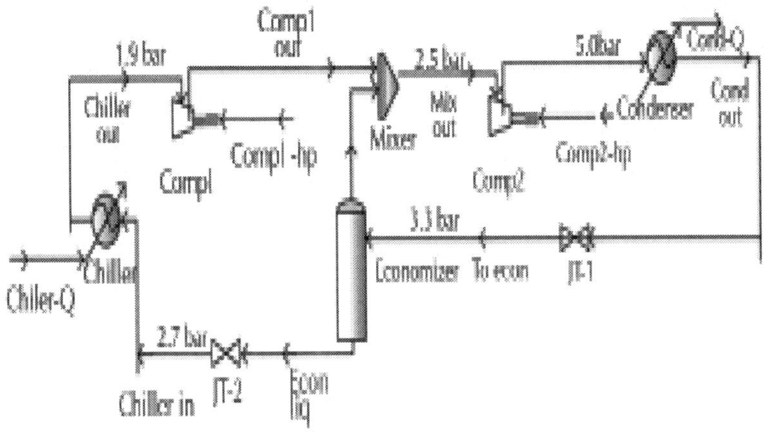

Figure 1: Process flow diagram of propane refrigeration of natural gas with application of Hysys software.

Expansion Turbine Unit

A turbo expander or an expansion turbine is a centrifugal or axial flow turbine through which a high pressure gas is expanded to produce useful work that is used to drive a compressor. Because work is extracted from the expanding high pressure gas, the expansion is an approximation by an isentropic process (a constant entropy process).

In the process of producing work, the expander lowers the bulk stream temperature which could result in partial liquefaction of the bulk stream (natural gas).

A systematic diagram of an expansion turbine (expander) is presented in Figure 2. The feed gas enters the gas treating unit after which it is cooled firstly in a gas-to-gas heat exchanger and secondly the propane chiller. The condensed liquid is removed in a separator and the gas from the separator is cooled further in the low temperature gas-to-gas heat exchanger and fed into a second cooled separator. Gas from the cooled separator expands through the expansion turbine to the pressure at the top of the demethanizer which varies from 100 to 450 bar. During the expansion, condensate is formed and the expander lowers the pressure of the inlet gas value to the demethanizer pressure of the range 100–450 bar.

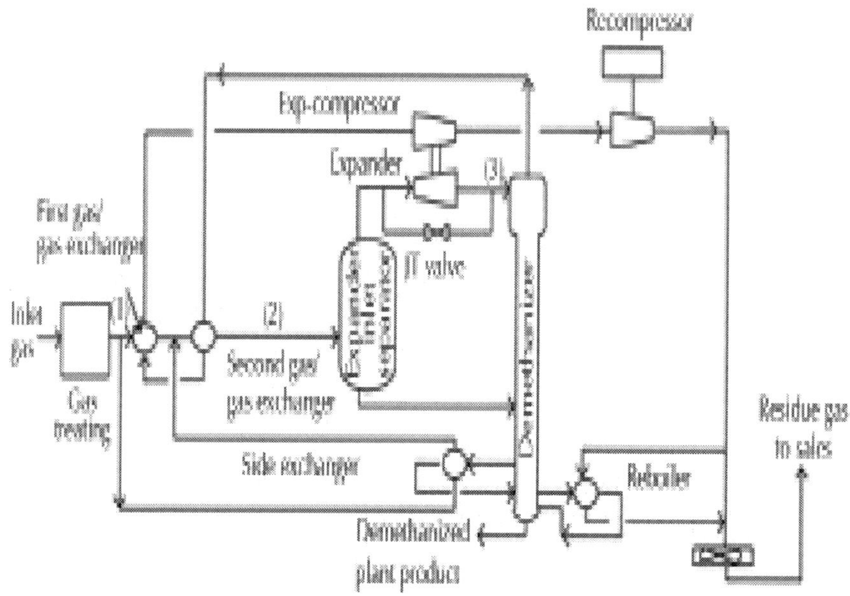

Figure 2: Flow diagram of expander process.

Thermodynamics Performance Parameters of Propane Refrigeration Unit: The relevant parameters required for thermodynamics performance evaluation of propane refrigeration of natural gas may be considered as follows.

Coefficient of Performance: This is defined as the ratio of the heat absorbed (Q_{in}) by the refrigerant while passing through the chiller to the work input (w_{in}) required compressing the refrigerant in the compressor. This is mathematically expressed as [9, 10]

$$\mathrm{COP} = \frac{Q_{\mathrm{in}}}{W_{\mathrm{in}}}, \tag{1}$$

But

$$W_{in} = W_{comp1} + W_{comp2}$$

$$= (1 - X_G)(H_B - H_A) + M_C(H_E - H_{D,MIX}),$$

$$COP = \frac{(1 - X_G) M_C (H_A - H_I)}{(1 - X_G)(H_B - H_A) + M_C(H_E - H_{D,MIX})}. \quad (2)$$

Refrigeration Capacity: The refrigeration capacity determines the rate of circulation of the refrigerant which, in turn, determines the design and size of various units such as condenser, compressor, evaporator (chiller), and expansion valves. This is expressed by the following equation:

$$R_c = \frac{M_C(H_A - H_I)}{210} \text{ KJ/ min}. \quad (3)$$

The mass flow rate of cooling water is calculated as follows:

$$M_{CW} = \frac{Q}{Cp_{CW}\Delta T}. \quad (4)$$

The overall performance of a compressor is affected mainly by the inlet pressure and the interstage cooler efficiency.

The simulation of process units in propane refrigeration resulted in loss of energy on the system. The second availability balance was applied to calculate the energy losses or lost work due to irreversibility of the process. The thermodynamics equations applied on each of the process units are as follows [10].

Compressor unit:

$$L_W = M_C\left[(H_A - T_o S_A)\right] - (H_B - T_o S_B) - W_E; \tag{5}$$

condenser unit:

$$L_W = T_o\left[M_C\left(S_C - S_B\right)\right] + M_{CW}S_{CW(out)} - S_{CW(in)},$$

$$M_{CW} = M_C\left[\frac{H_E - H_F}{H_{CW(out)} - H_{CW(in)}}\right]; \tag{6}$$

expansion valve:

$$L_W = M_C T_o\left(S_B - S_C\right),$$

$$L_W = M_C T_o\left[\left(S_A - S_D\right) - \left(\frac{H_A - H_D}{T_i}\right)\right]. \tag{7}$$

Refrigerant chiller: the chiller is the unit where the process gas (natural gas) gives up its energy to the liquid refrigerant. The refrigerating effect which is the amount of heat absorbed by refrigerant or removed from the refrigerated space is expressed as

$$Q_{in} = \left(1 - X_G\right) M_C\left(H_A - H_I\right). \tag{8}$$

The fraction of refrigerant vaporized in the chiller is determined from enthalpy balance as follows:

$$M_C(H_H)_L = M_C X_V\left(H_I\right) + M_C\left(1 - X_v\right)\left(H_I\right)_L, \tag{9}$$

Where X_v is the mole fraction of refrigerant that evaporates as it throttles to the economizer unit.

Expansion Turbine Unit. The relevant thermodynamics parameters applied in expansion turbine unit in a gas plant are as follows.

- *Isentropic Turbine Efficiency.* This is defined as the ratio of turbine actual work output to the work output that would be achieved if the process between the inlet and outlet pressure was isentropic. By definition [11],

$$\eta_E = \frac{h_1 - h_{2a}}{h_1 - h_{2,S}}. \tag{10}$$

- *Isentropic and Actual Work.* The isentropic work of a turbine is the work done by the turbine during the constant entropy whereas actual work is the gross work of a turbine:

$$W_{\text{act}} = m\left(h_2 - h_1\right),$$
$$W_{\text{is}} = -m\left(h_1 - h_{2,S}\right). \tag{11}$$

- *Power Recovery.* The expander gas power recovery is the output power of the turbine list to drive the compressor. By definition,

$$P_g = -\eta_m\left(h_1 - h_{2,S}\right). \tag{12}$$

- *Theoretical and Actual Discharge Temperature.* The discharge and actual temperatures of a turbine are expressed by the following equations [11, 12]:

$$T_2' = T_1\left(\frac{p_2}{p_1}\right)^{(k-1)/k},$$

$$T_2 = T_1 + T_1\left[\left(\frac{p_2}{p_1}\right)^{(K-1)/K} - 1\right]\eta_E. \tag{13}$$

The results obtained from the evaluation with the above equations are shown in graphical form in Figures 3and 4.

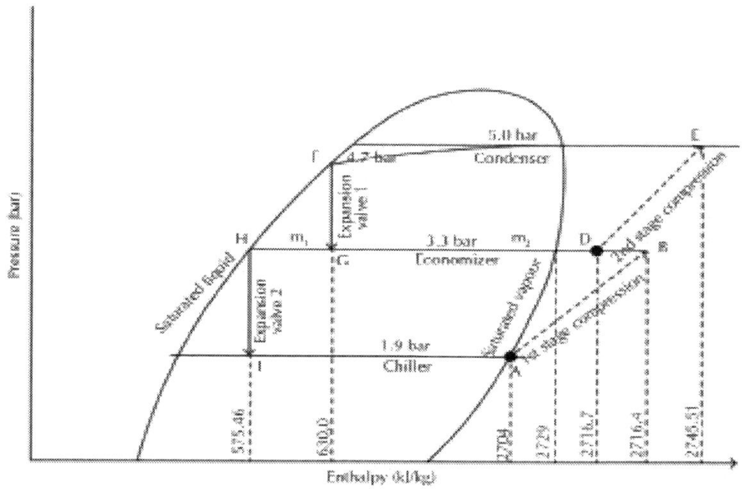

Figure 3: Pressure enthalpy diagram of a propane refrigeration cycle.

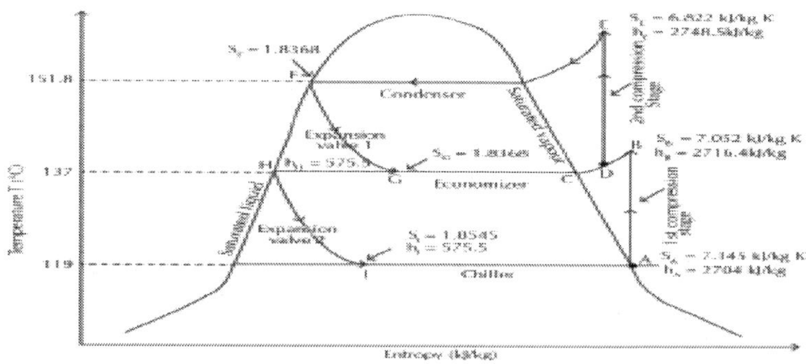

Figure 4: Temperature entropy diagram of a propane refrigeration cycle.

RESULTS AND DISCUSSION

The results on the refrigeration unit of a gas plant were obtained using the P-S and T-S diagrams.

The two-stage compression refrigeration system in Figures 3 and 4 operates with a pressure range of 1.9 and 5.0 bar. The refrigerant enters the first stage compressor at a suction pressure of 1.9 bar with enthalpy and entropy values of 2704 KJ/kg and 7.145 KJ/kg·K, respectively. It discharges into a mixer where it mixes up with the vapor leaving the top of the flash economizer. The refrigerant is compressed to the second stage compressor operating with a suction pressure of 2.5 bar with enthalpy and entropy values of 2716.4 KJ/kg and 7.052 KJ/kg·K, respectively, and discharges at the condenser inlet pressure of 5.0 bar, where −17,386.7 J/kg of heat is rejected to the surrounding. The enthalpy and entropy values of the condenser as a saturated vapour at a pressure of 4.7 bar were observed to be 630 KJ/kg and 1.8363 KJ/kg·K, respectively. The refrigerant is throttled to a flash economizer at an intermediate pressure of 3.3 bar. During flashing 0.0034 kg/s of vapor evaporates from the economizer drum while 0.1305 kg/s of liquid is throttled to the chiller at a pressure of 1.9 bar with enthalpy and entropy values of 575.5 KJ/kg and 1.8545 KJ/kg·K, respectively. The process gas (natural gas) countercurrently exchanges heat and gives up its energy to the liquid refrigerant at a temperature of 41°C. The result of the two-stage compression refrigeration is summarized in Table 1.

Table 1: Summary of two-stage compression refrigeration

Parameters	Units	1st stage compressor	2nd stage compressor
Suction pressure	bar	1.9	2.5
Discharge pressure	bar	2.5	5.0
Compression ratio	—	1.316	1.414
Isentropic work of compression	J/kg	20,644.25	22,094.91
Actual work of compression	J/kg	20,644.25	21,872.85
Isentropic efficiency	%	48.56	51.98

Mechanical efficiency	%	20.71	46.51
Volumetric efficiency	%	55.0	46.7
Actual power requirement	KW	0.0969	0.0960
Cooling water consumption in intercoolers	m3/s	7.2612	19.282
Compressor capacity	m3/s	2.1566	2.1096
Work input in comp 1 and comp 2	J/kg	273.14	273.14
Work output	J/kg	1197.89	2414.11

The process parameters of other auxiliary units in propane refrigeration plant are shown in Table 2.

Table 2: Process parameters of other auxilliary units

Parameters	Unit	Condenser	Economizer	Expansion valve 1	Expansion valve 2	Refrigerant chiller
Suction pressure	bar	5.0	3.3	4.7	3.3	2.3
Discharge pressure	bar	4.7	—	3.3	1.9	1.9
Heat load	J/kg	−17,386.7	5733.78	—	—	17,036.04
Mass flow rate of vapor—economizer	kg/s	—	0.0034	—	—	—
Mass flow rate of liquid—economizer	kg/s	—	0.1305	—	—	—
Mass flow rate of cooling water—condenser	kg/s	6.72	—	—	—	—
Joule Thompson coefficient	—	—	—	8.57	12.86	—
Mole fraction of liquid remaining in chiller	—				0.035	0.00014

Mole fraction of vapor evaporated in chiller	—		0.025	0.025	—	0.045
Mass flow rate of liquid—condenser accumulator	kg/s	0.1339				

The energy balance of the propane refrigeration unit obtained from the evaluation based on the equations highlighted for the analysis is shown in Table 3.

Table 3: Energy balance propane refrigeration unit

Process units	Energy gain (KJ/kg)	Energy lost (KJ/kg)	% of lost work
Compressor 1	12.30	−334.28	−0.015
Compressor 2	32.11	−501.45	−0.023
Condenser	—	12483.89	0.06
Expansion valve JT-1	—	12495.54	0.06
Expansion valve JT-2	—	−1435547.8	6.62
Economizer drum	5733.78	—	—
Refrigerant chiller	17036.04	−1242710.43	−5.73
Total	22814.23	216981.07	0.972

By the design conditions, the centrifugal compressors used in compressing the refrigerant should operate with a pressure ratio of 1.2 : 1 and 1.4 : 1, isentropic efficiency of 70–80%, volumetric efficiency of 60–89%, and mechanical efficiency of 20–50%. By analytical method, the pressure ratio was found to be 1.316 and 1.414; the isentropic efficiency was found to be 48.56% and 51.97%; the volumetric efficiency was found to be 55.0% and 46.7% for the first and second stage compressors, respectively, while the mechanical efficiencies are 57.7% and 46.51%, respectively. The refrigeration cycle was operating with an overall coefficient of performance of 62.37 with a refrigerating capacity of 4,922.2 tons after examining the performance of other auxiliary units within the systems. Tables 1, 2, and 3 show values obtained from the evaluation. The Kelvin-Planck statement of the second law of thermodynamics tells us that it is impossible to have a heat engine that will convert all the heat received from the high temperature source, Q_H, into useful work in a thermodynamic cycle. It is necessary to reject part of the received heat to the low temperature source, Q_L. In other words, it is impossible to have a 100% efficiency heat engine as corroborated by Kumar et al. [13] and José and Simões-Moreira [14].

CONCLUSIONS

The thermodynamics equations were applied to construct Figures 3 and 4 which determine the performance of the plant. The recorded efficiencies ranging from 63.92% to 77% have shown that the overall performance has deviated from the design efficiency. This condition may increase the operational cost of the plant.

The following recommendations are highlighted to ensure optimum efficiency and reliability of NGL plant.

- The feed gas must be free from CO_2 and water. This affects plant efficiency and operations if not properly checked by freezing the fittings, valves, and other associated equipment.
- The inlet strainer differential pressure must not be high; otherwise the expander will trip off on high differential pressure.
- The refrigeration unit must be operated within the operating and design conditions to avoid overfreezing or warming of the demethanizer column.
- There is need for proper insulation of piping system within the NGL extraction plant. The reason is not to allow the surrounding heat to enter the system, thereby warming the system, and there may be freezing out of the plant.
- Proper sizing of the process line using the various line balance methods to compare the amount of natural gas entering the

process plant with the amount put in should be adopted at the design stage of the process plant.

REFERENCES

1. M. M. Rahman, T. K. Ibrahim, K. Kadirgama, R. Mamat, and R. A. Bakar, "Influence of operation conditions and ambient temperature on performance of gas turbine power plant," Advanced Materials Research, vol. 189–193, pp. 3007–3013, 2011.
2. H. Taniguchi and S. Miyamae, "Power generation analysis for high-temperature gas turbine in thermodynamic process," Journal of Propulsion and Power, vol. 16, no. 4, pp. 557–561, 2000.
3. A. Khaliq and S. C. Kaushik, "Thermodynamic performance evaluation of combustion gas turbine cogeneration system with reheat," Applied Thermal Engineering, vol. 24, no. 13, pp. 1785–1795, 2004.
4. A. B. Keith and R. H. Kenneth, Economic optimization of natural gas processing plant including business aspects [Ph.D. dissertation], Texas A & M University, 2006.
5. K. L. Jibril, A. I. Al-Humaizi, A. A. Idriss, and A. A. Ibrahi, "Simulation of turbo-expander processes for recovering of natural gas liquids from natural gas," Saudi Aramco Journal of Technology, pp. 9–14, 2005.
6. S. T. Donnelly, J. C. Polasek, and J. A. Bullin, Process Simulation and Optimization of Cryogenic Operations Using Multi-Stream Brazed Aluminium Exchangers, Bryan Research and Engineering, 2006.
7. T. Ganapathy, N. Alagumurthi, R. P. Gakkhar, and K. Murugesan, "Exergy analysis of operating lignite fired thermal power plant," Journal of Engineering Science and Technology Review, vol. 2, no. 1, pp. 123–130, 2009.
8. A. A. Ujile and L. Alawa, "Thermodynamic evaluation of a natural gas processing plant: case study—Obiafu/Obrikom, Rivers State, Nigeria," Journal of Advanced Science and Engineering Technology, vol. 2, no. 1, pp. 76–80, 2012.

9. T. D. Eastop and A. McConkey, Applied Thermodynamics for Engineering Technologists, Pearson Education, 5th edition, 2005.
10. Y. A. Cengel and M. A. Boles, Thermodynamics, an Engineering Approach, McGraw-Hill, New York, NY, USA, 4th edition, 2002.
11. I. S. Stanley, Chemical, Biochemical, and Engineering Thermodynamics, John Wiley & Sons, 4th edition, 2006.
12. C. Arora, Refrigeration and Air Conditioning, McGraw Hill, Delhi, India, 5th edition, 2005.
13. A. Kumar, S. S. Kachhwaha, and R. S. Mishra, "Thermodynamic analysis of a regenerative gas turbine cogeneration plant," Journal of Scientific and Industrial Research, vol. 69, no. 3, pp. 225–231, 2010.
14. R. José and Simões-Moreira, "Thermal Power Plant Performance Analysis," 2012,http://www.springer.com/978-1-4471-2308-8.

Chapter 2

Geomechanical Properties of Unconventional Shale Reservoirs

Mohammad O. Eshkalak[1], Shahab D. Mohaghegh[2], and Soodabeh Esmaili[3]

[1]Department of Petroleum and Geosystems Engineering, the University of Texas at Austin, Austin, TX 78712, USA

[2]Department of Petroleum and Natural Gas Engineering, West Virginia University, Morgantown, WV 26506, USA

[3]Occidental Petroleum Corporation, Bakersfield, CA 93311, USA

ABSTRACT

Production from unconventional reservoirs has gained an increased attention among operators in North America during past years and is believed to secure the energy demand for next decades. Economic production from unconventional reservoirs is mainly attributed to realizing the complexities and key fundamentals of reservoir formation properties. Geomechanical well logs (including well logs such as total minimum horizontal stress, Poisson's ratio, and Young, shear, and bulk modulus) are secured source to obtain these substantial shale rock properties. However, running these geomechanical well logs for the entire asset is not a common practice that is associated with the cost of obtaining these well logs. In this study, synthetic geomechanical well logs for a Marcellus shale asset located in southern Pennsylvania are generated using data-driven modeling. Full-field geomechanical distributions (map and volumes) of this asset for five geomechanical properties are also created using general geostatistical methods coupled with data-driven modeling. The results showed that synthetic geomechanical well logs and real field logs fall into each other when the input dataset has not seen the real field well logs. Geomechanical distributions of the Marcellus shale improved significantly when full-field data is incorporated in the geostatistical calculations.

INTRODUCTION

Shale gas reservoirs, which are also called source rock reservoirs (SRR), have some unique attributes that make hydraulic fracturing an essential option in order to commence an economic level of the natural gas production. Unlike conventional gas reservoirs, insufficient permeability, ultra-low porosity of shale rock, and limited reservoir contact area, but vastly organic-rich formation, cannot offer production in a commercial value without stimulation processes. Many studies are conducted from shale pore-scale level to field scale reservoir simulations to improve the understanding of complex flow behavior that are developed and discussed through numerical, analytical, and semianalytical reservoir models for unconventional reservoirs [1–10]. However, in order to predict the performance of a shale gas reservoir, implementing accurate shale rock properties is essential for developing a geologic model for the entire asset. Hence, it is very critical to access more data while working on an unconventional reservoir. In this study, synthetic data are generated using artificial intelligence and data mining techniques (AI&DM).

Principal stress profile of an oil and gas reservoir depends highly on the rock geomechanical properties. Geomechanical properties of reservoir rock include Poisson's ratio, total minimum horizontal stress, and bulk, Young, and shear modulus. These properties play significant role in current development plans of shale assets compared to conventional reservoirs that have established sufficient information available. Moreover, having access to geomechanical data can assist engineers and geoscientists during geomechanical modeling, hydraulic fracture treatment design, and reservoir simulation in shale gas fields across the U.S. and worldwide. Geomechanical well logs are one of the sources that secure such data. Running geomechanical well logs (in all wells in a shale asset) is not common practice among operators and the reason is attributed to the cost associated with running such logs. In this paper, data-driven models are developed to accurately determine the Marcellus shale rock properties.

Artificial intelligence and data mining have been used within last 20 years in reservoir modeling and characterization to perform analysis on formation of interest [11, 12]. Also, some studies indicated that artificial neural network (ANN) is a powerful tool for pattern recognition and system identification such as methodology developed by Mohaghegh et al. in 1998 [13] to generate synthetic magnetic resonance imaging (MRI) logs using conventional logs such as SP, GR, and resistivity. Their methodology incorporated an artificial neural network as its main tool to generate the target variable. The synthetic magnetic resonance imaging logs were generated with a high degree of accuracy even when the model developed used data not employed during model development. Moreover, Basheer and Najjar demonstrated that ANN is suitable to predict and classify soil compaction and rocks characteristics as well as determining some mechanical parameters such as Young's modulus, Poisson's ratio [14]. They mainly investigated the neural network capability in solving geotechnical engineering problems and they provided a general view of some neural network application in their field of research.

In this study, it is demonstrated that AI&DM technology is able to develop data-driven models for generating rock geomechanical properties. The overall work-flow includes development of synthetic geomechanical well logs from conventional logs such as gamma ray and bulk density that are commonly available. These data-driven models used around 30 percent of data (coming from geomechanical logs) of the entire asset, which were available to expand them for the rest of the field with conventional logs but no geomechanical logs. Data-driven models have been validated with blind wells. Blind wells are wells with actual data which are selected due to different locations in the asset of 100 horizontal gas wells. Moreover, the logs generated from data-driven models are used to build an integrated field-wide geomechanical distribution (maps and volumes) for rock geomechanical properties. In this work, the ultimate purpose is meant to propose a technique in order to omit the necessity of running geomechanical logs for the entire asset once such logs are obtained for some portion of the asset to be used in the data-driven models. The number of well logs required to run the data-driven models that leads to an accurate result is determined to be 30 percent of the wells in an unconventional asset. In this study, just 30 out of 100 horizontal wells in the Marcellus shale asset in southern Pennsylvania own actual geomechanical well logs that are provided by a major service company.

METHODOLOGY

In order to accomplish the objectives of this study, a methodology and thorough procedure are required to be defined. The methodology used to accomplish the objectives of this study includes four steps as indicated below. A detailed description of each step is explained afterwards:

- data preparation,
- data-driven model development using AI&DM,
- validation of data-driven models,
- geomechanical property distribution.

Step (a): Data Preparation

Data preparation is the most important step in developing data-driven models due to the fact that all the other steps are using the data prepared in this step. Data preparation involves checking the data for accuracy, entering data in a right format in a computer file, and developing and documenting a database structure that integrated the various properties used in the next steps. In this study, a dataset form the available well logs are required to be prepared that portrays specific property of the rock versus depth. First, production pay zone must be identified from the conventional well logs along with the horizontal wells trajectory. After identifying the depth of the producing zones for Marcellus shale from conventional well logs such as gamma ray, available information of each individual well is extracted from well logs in every one or half a foot according to its log characteristics and tools used. Also, in order for the models to understand the differences between production pay zones and while using these datasets, it is essential to specify the contrast between pay zones (upper Marcellus, Purcell, lower Marcellus, and Onondaga) and the adjacent rocks, nonshale. To account for this contrast, 50 feet depth of log data located above and below the pay zones of interest is also added to the main dataset.

The prepared dataset contains rows and columns consisting of following data that are recorded versus depth: the wells name, the well coordinates, the values for gamma ray (GR), bulk density (BD), sonic porosity, bulk modulus (BM), shear modulus (SM), Young's modulus (YM), Poisson's ratio (PR), and total minimum horizontal stress (TMHS) for each horizontal shale well. It must be emphasized that not all the wells include geomechanical well logs; thus geomechanical values are only recorded for the wells that have such real data, 30 wells.

Step (b): Data-Driven Models Development

In this step, the prepared dataset was processed using backpropagation algorithm of neural network into two main parts.

Part 1 (Conventional Models): In order to have consistent conventional logs data for all wells for the shale production pay zone, part 1 is defined and the conventional logs are generated for those wells that missed some conventional logs. As it is shown in Figure 1, the bulk density and sonic porosity for 30 wells were produced by using two different data-driven models. First model (neural network model 1) used gamma ray, depth, location, and well coordinates as an input to develop training, calibration, and verification segments for generating the bulk density of around 30 wells in the asset where bulk density and sonic data were missing. Second model (neural network model 2) used also bulk density as input beside inputs of first step to generate sonic porosity for the part of asset without this property (around 30 wells also used in this part).

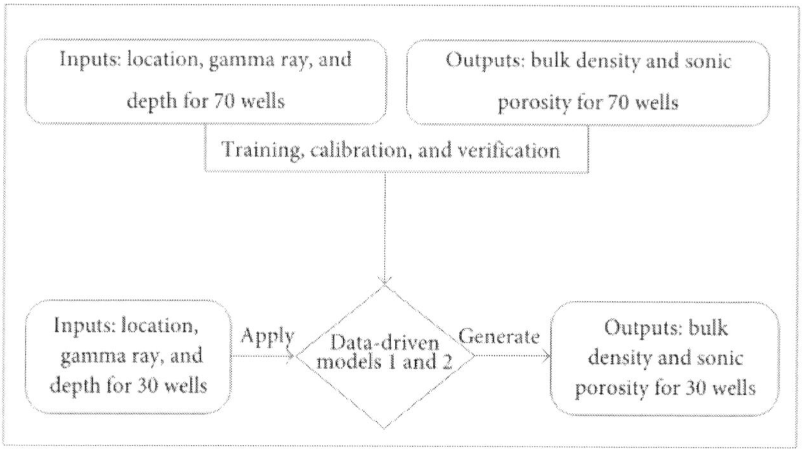

Figure 1: Part 1—data-driven model for conventional logs.

At the end of this step, all existing wells in the asset have the required conventional well log properties to be used in part 2.

Part 2 (Geomechanical Models): After completing the missing data for conventional logs from part 1 for all horizontal wells, another neural network structure is used to develop five different data-driven models (models 3 to 7) as shown in Figure 2 to generate the geomechanical well logs for all wells. As it is shown in Figure 2, this step consists of five neural network models in which the inputs were completed in each step by using the generated geomechanical property in the previous step. In detail, for each of these neural networks, the same conventional logs of 30 wells have been provided as the input and one geomechanical property was generated at a time. Then each generated geomechanical property was used as an input for the next neural network model and the process continued until all five geomechanical properties of interest were generated for the entire Marcellus shale asset.

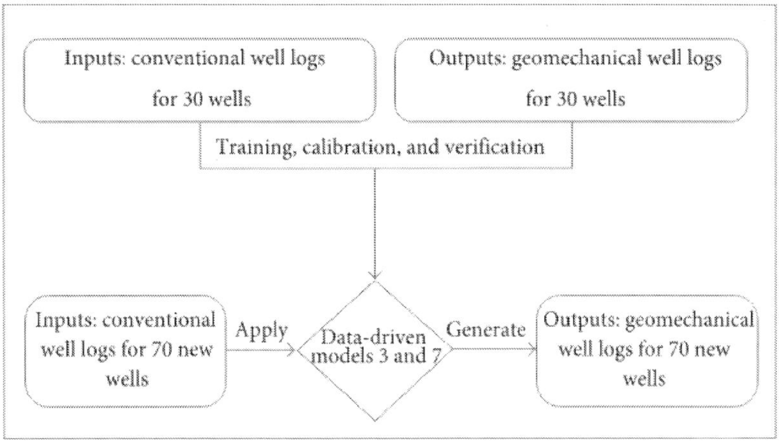

Figure 2: Part 2—data-driven models for geomechanical logs.

All models are multilayer networks that are trained using a backpropagation method of neural network technology. In order to achieve the least error in backpropagation method, a different percentage of datasets are incorporated in the models and finally it is concluded that considering 80% of data used in the training process, and by considering 20% for the calibration process and the remaining for the last step, verification (10% for each), the universal backpropagation error is minimized. A general backpropagation scheme and formulation is explained in Appendix. In the next section, two different methods for error calculation S are explained.

Error Analysis: The mean absolute percentage error (MAPE), also known as mean absolute percentage deviation, is a measure of accuracy of a method for constructing fitted time series values in statistics, specifically in trend estimation. It usually expresses accuracy as a percentage and is defined by

$$\text{MAPE} = \frac{100\%}{n}\sum_{t=1}^{n}\left|\frac{At - Ft}{At}\right|, \qquad (1)$$

Where At is the actual value and Ft is the forecast value.

The difference between At and Ft is divided by the actual value At again. The absolute value in this calculation is summed for every fitted or forecasted point in time and divided again by the number of fitted points n. Multiplying by 100 makes it a percentage error. Also, R-squared is defined and determined in (2) for all three steps, training, calibration, and verification. The higher the R-squared, the closest the results to the actual values:

$$R\text{-squared} = 1 - \frac{\text{Sum of squared distances between the actual and predicted values}}{\text{Sum of squared distances between the actual and their mean values}}. \quad (2)$$

In our study the highest achieved R-squared was around 98 percent and the lower one in some cases around 89 percent, and in both situations, the results presented are highly acceptable. A higher level of R-squared reflects, in all three stages of training, calibration, and verification, a reliable correlation between actual and generated data. It is also important to mention that during the initial training of datasets; the results obtained were with low R-squared. Unsuccessful behavior of models was understood because of having some wells with log data for each 0.5 ft., which is in contrast with the rest of the wells with every available 1 ft. log data. Once piece of data of 0.5 ft. turned to 1 ft., which is considered as discrepancy that misleads the data-driven models; the results came out properly and the data-driven models showed rapid improvements. Further, second issue that resulted in smoother behavior of the data-driven models was related to the removal of nonshaly thin intervals log data from the pay zones that exists within the upper Marcellus, Purcell, lower Marcellus, and Onondaga. Once these layers were removed, the models converged much faster with very low backpropagation error.

Step (c): Data-Driven Model Validation

This step explains a robust method to analyze the accuracy and validity of the data-driven model's results, although the universal error of all the models was very low according to previous section. To examine the models validity, the well log data of some wells (which have both real conventional and geomechanical logs) was removed from the training dataset and it was attempted to regenerate the geomechanical logs. These removed wells are so called blind wells. Blind wells have been chosen from different location in the Marcellus shale asset under study. Then, data-driven models used this new dataset to be trained to generate geomechanical properties for blind wells. Data-driven models number 3 to 7 was separately validated to generate geomechanical properties.

In each step, the generated property compared and plotted against the actual values which had been removed from main dataset. The results of this step are explained in Results and Discussions section.

Figures 4 through 8 demonstrate the actual well logs and generated logs for 5 blind wells that are chosen form different location in the asset as illustrated in Figure 3. To compare the results, both actual and generated properties are plotted in the same figure like an actual well log. Properties such as bulk modulus, Young modulus, Poisson's ratio, shear modulus, and total minimum horizontal stress are presented, respectively. Blue line shows the actual value and the red line is for generated values by data-driven models. For well #1 to well #4, there is a perfect match between blue and red lines that shows the models have generated the exact actual data. These wells are in proximity of wells with actual geomechanical properties according to their locations and depths. As it was expected, results shown for these wells are accurate which demonstrate data-driven model's capability and accuracy in generating geomechanical properties of Marcellus shale.

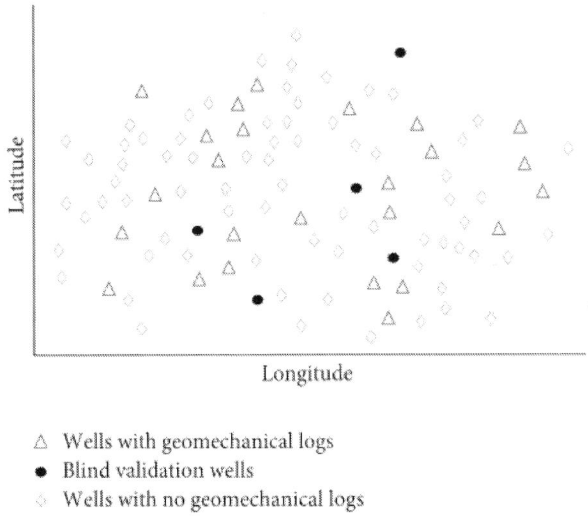

Figure 3: Marcellus shale gas field, southern Pennsylvania.

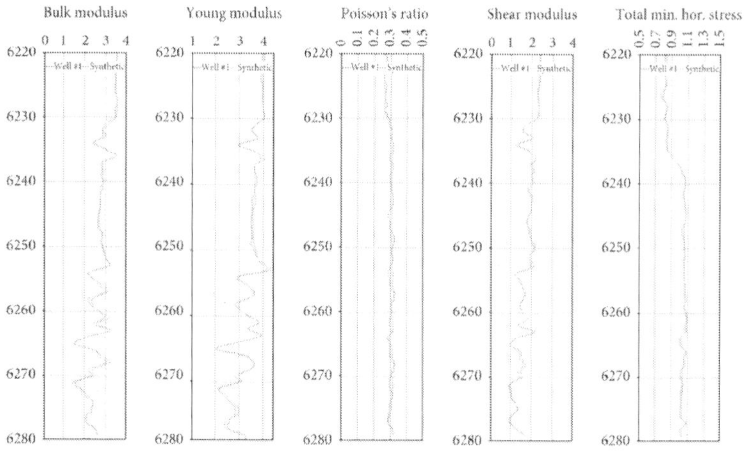

Figure 4: Well #1, actual versus generated well logs.

Step (d): Geomechanical Property Distribution

The first objective of this paper is accomplished in the previous section and the geomechanical properties are generated for all existing wells in the Marcellus shale asset. To accomplish this step, different geostatistical methods, from Petrel commercial software, are considered to create geomechanical property distribution for the Marcellus shale field. Further, geomechanical well logs generated from the data-driven models are coupled with a commercial reservoir simulator in order to create geomechanical distributions for properties such as total minimum horizontal stress, Poisson's ratio, and Young, shear, and bulk modulus.

Sequential Gaussian simulation (SGS) is finally used to create distribution according to well locations for the entire field due to its very smooth and consistent surfaces and distributions (maps) obtained compared to other methods. Two types of maps were created. First map is only incorporated with 30 wells which already had actual geomechanical logs. The second map is related to entire field (70 wells with generated property and 30 wells with actual data). With comparing these two maps, significant difference between geomechanical property distribution with and without having full-field data is observed as shown in Figures 9 through 13. Ten maps that show distribution of five rock geomechanical properties in the Marcellus shale asset were created.

RESULTS AND DISCUSSIONS

The Marcellus shale under study consists of 100 multifractured horizontal wells. Figure 3 depicts the distribution of existing wells in the asset that is used in this study. Table 1 shows the information and number of wells that were used to develop data-driven models in different steps as well as the validation purpose in step (c).

Table 1: Information used for databases for developing data-driven models

Well identifier	Description	Conventional well logs	Geomechanical well logs	Number of wells
Circle	Blind validation wells	Yes	Yes	5
Triangle	Wells used for validation	Yes	Yes	25
Diamond	Wells with no geomechanical logs	Yes	No	70

In this study, a multilayer neural networks or multilayer perceptions are considered to develop the data-driven models. These networks are most suitable for pattern recognition specially in nonlinear problems neural network that have one hidden layer with different number of hidden neurons that are selected based on the number of data records available and the number of input parameters selected in each training process.

The training process of the neural networks is conducted using a backpropagation technique. In the training process, the dataset is partitioned into three separate segments. This is done in order to make sure that the neural network will not be trapped in the memorization phase. Moreover, the intelligent partitioning process allows the network to adapt to new data once it is being trained. The first segment, which includes the majority of the data, is used to train the model. In order to prevent the memorizing and overtraining effect in the neural network training process, a second segment of the data is taken for calibration that is blind to the neural network and at each step of training process, the network is tested for this set. If the updated network gives better predictions for the calibration set, it will replace the previous neural network; otherwise, the previous network is selected. Training will be continued once the error of predictions for both the calibration and training dataset is satisfactory. This will be achieved only if the calibration and training partitions are showing similar statistical characteristics. Verification partition is the third and last segment used for the process that is kept out of training and calibration process and is used only to test the precision of the neural networks. Once the network is trained and calibrated, then the final model is applied to the verification set. If the results are satisfactory then the neural network is accepted as part of the entire prediction system [15, 16].

Figures 4 to 8 show the actual well logs and generated logs for 5 blind wells shown as black circles in Figure 3. To compare the results, both actual and generated properties are plotted in the same figure similar to an actual well log. In these plots, properties such as bulk modulus, Young modulus, Poisson's ratio, shear modulus, and total minimum horizontal stress are presented, respectively. Blue line shows the actual value and the red line is for generated values by data-driven models. For well #1 to well #4 (Figures 5, 6, and 7), there is perfect match between blue and red lines. These wells are in proximity of wells with actual geomechanical properties according to their locations and depths. As it was expected, results shown for these wells are accurate which demonstrate data-driven models capability in predicting geomechanical properties.

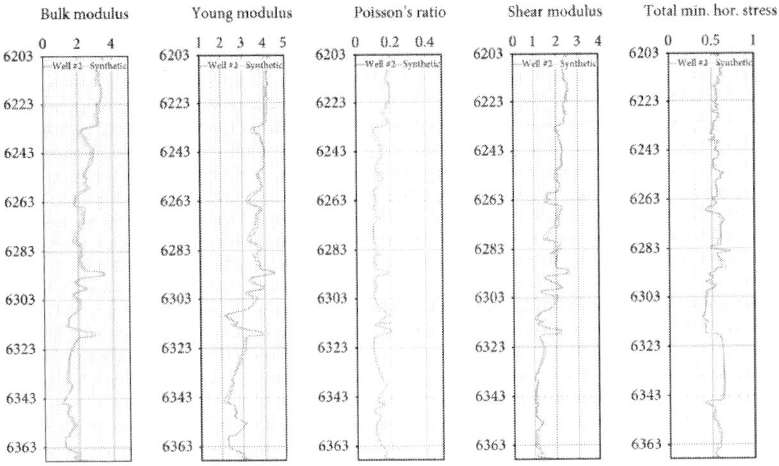

Figure 5: Well #2, actual versus generated well logs.

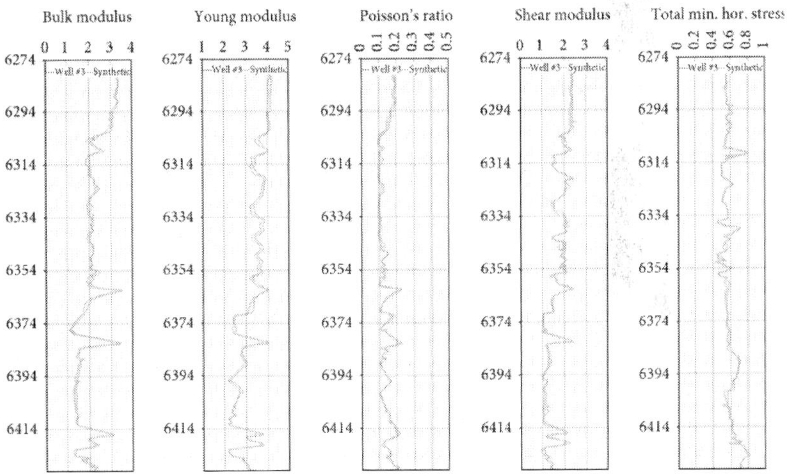

Figure 6: Well #3, actual versus generated well logs.

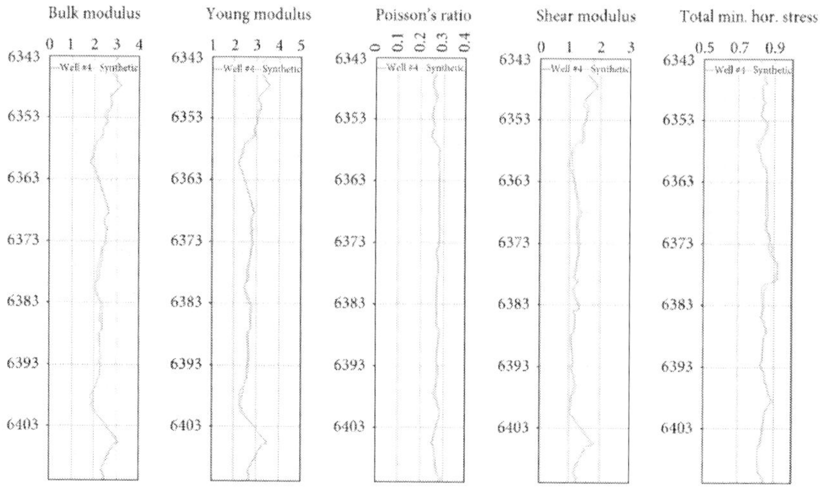

Figure 7: Well #4, actual versus generated well logs.

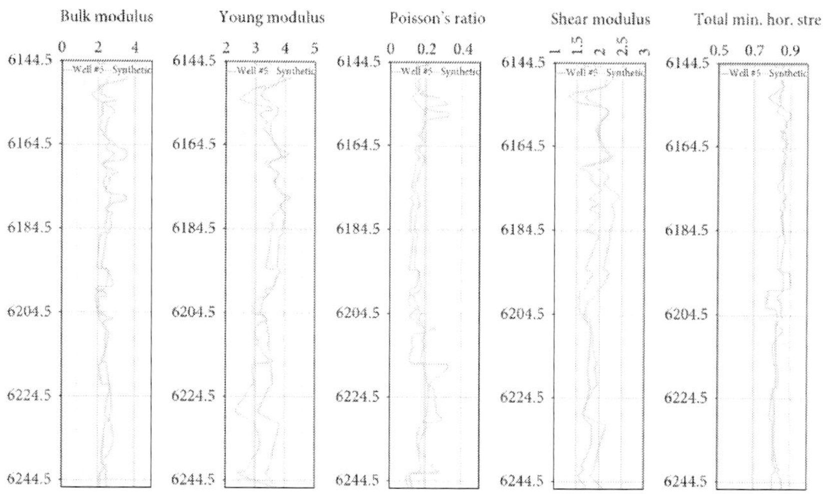

Figure 8: Well #5, actual versus generated well logs.

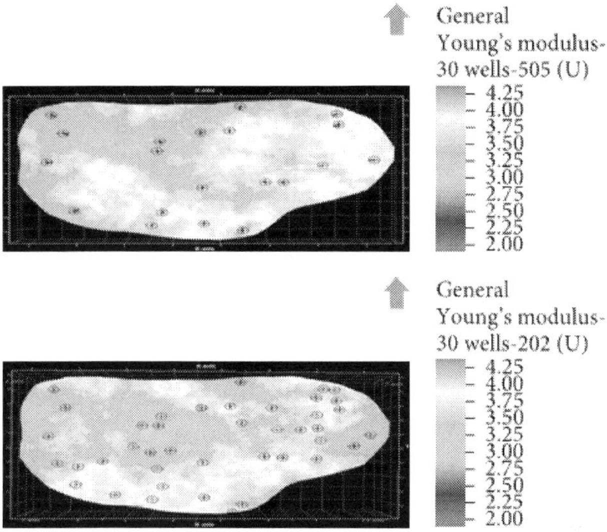

Figure 9: Young modulus.

For well #5 in Figure 8, the generated data is not in agreement with the actual logs and it is because of the location of the well that is far from (upper side of the asset in the field—Figure 2) the rest of wells in the asset that we have used for the training purposes. This fact indicates that the models could not predict the behavior of outlier wells and most importantly emphasizes on the fact that data-driven modeling is perfect for interpolations and not accurate for the extrapolation as it is in agreement with the neural network literature. Moreover, it is found that the depth of producing pay zone of this well #5, compared to other four blind wells, is different (out of range) and it might be another reason related to the fact that models could not capture the behaviors very well.

Figures 9, 10, 11, 12, and 13 are showing distributions and maps for the five geomechanical rock properties of interest in this study. For each property, there are two distributions: one that is generated by using the actual data and the second that considered the information of both generated and actual data (full-field data). A comparison between maps for each property demonstrates that more reasonable and accurate distribution is achieved using more data for the asset. The sequential Gaussian simulation (SGS) algorithm was used in order to generate these maps. In the top distribution map, plus signs represent the wells with actual data which have been used in dataset for training, calibration, and verification during data-driven model development.

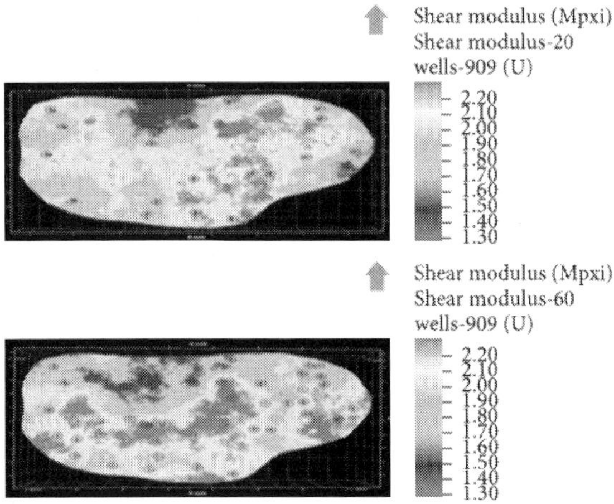

Figure 10: Shear modulus.

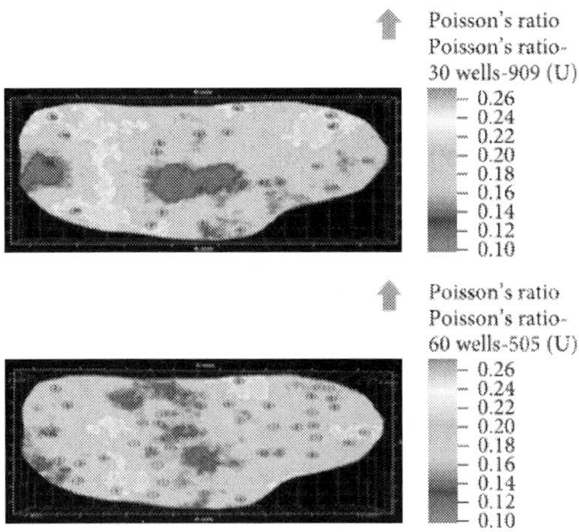

Figure 11: Poisson's ratio.

Figure 12: Total minimum horizontal stress.

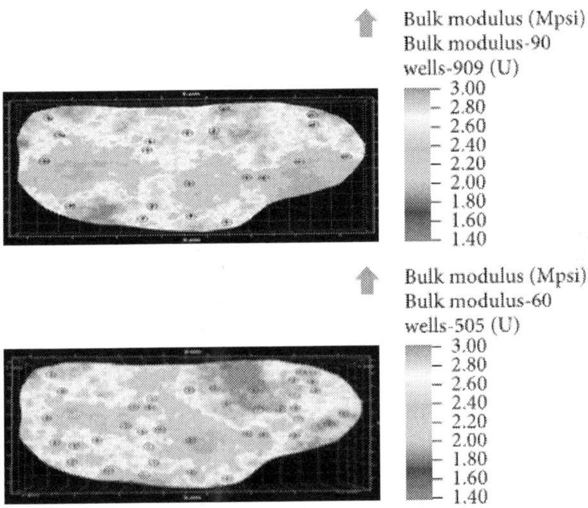

Figure 13: Bulk modulus.

CONCLUSIONS

In this study, it is demonstrated that the data-driven modeling using AI&DM technology is a reliable and robust tool to obtain accurate results for generating synthetic geomechanical logs for unconventional shale resources. In simple terms, we used conventional well logs to generate field-wide geomechanical properties and distribution maps of geomechanical properties for the entire asset, Marcellus shale in southern Pennsylvania.

Five data-driven models were designed, trained, and validated to predict five geomechanical properties of interest for Marcellus shale unconventional reservoir. First, data mining issue in this study, removing nonshaly intervals and adding 50 feet contrast zone, was successfully managed to lead a reliable prediction with least error calculated in backpropagation method. Also, second validation process, the use of 5 blind wells, was performed to show the robustness and accuracy of data-driven models for predicting Young modulus, Poisson ratio, bulk modulus, shear modulus, and total minimum horizontal stress.

Geomechanical property distribution maps of the entire asset illustrated a significant difference between distributions when there are just a few available pieces of actual data rather than having access to the full-field data. These synthetic geomechanical logs and property distributions for Marcellus shale exhibit a great deal of assistance to better performing reservoir modeling, characterization and the optimization of hydraulic fracturing issues related to the current Marcellus shale development process. Authors expect these models will conclude also accurate results in other unconventional shale resources.

REFERENCES

1. C. L. Cipolla, E. P. Lolon, J. C. Erdle, and B. Rubin, "Reservoir modeling in shale-gas reservoirs," SPE Reservoir Evaluation & Engineering, vol. 13, no. 4, pp. 638–653, 2010.
2. W. Yu and K. Sepehrnoori, "Simulation of gas desorption and geomechanics effects for unconventional gas reservoirs," in Proceedings of the SPE Western Regional/Pacific Section AAPG Joint Technical Conference: Energy and the Environment Working

Together for the Future, pp. 718–732, Monterey, Calif, USA, April 2013.

3. S. Esmaili, A. Kalantari-Dahaghi, and S. D. Mohaghegh, "Forecasting, sensitivity and economic analysis of hydrocarbon production from shale plays using artificial intelligence & data mining," in Proceedings of the Canadian Unconventional Resources Conference, Calgary, Canada, October-November 2012, paper SPE 162700.
4. T. W. Patzek, F. Male, and M. Marder, "Gas production in the Barnett Shale obeys a simple scaling theory," Proceedings of the National Academy of Sciences of the United States of America, vol. 110, no. 49, pp. 19731–19736, 2013. · ·
5. U. Aybar, M. O. Eshkalak, K. Sepehrnoori, and T. W. Patzek, "The effect of natural fracture›s closure on long-term gas production from unconventional resources," Journal of Natural Gas Science and Engineering, 2014. ·
6. M. O. Eshkalak, "Simulation study on the CO_2-driven enhanced gas recovery with sequestration versus the re-fracturing treatment of horizontal wells in the U.S. unconventional shale reservoirs," Journal of Natural Gas Science and Engineering, 2014. ·
7. M. O. Eshkalak, U. Aybar, and K. Sepehrnoori, "An integrated reservoir model for unconventional resources, coupling pressure dependent phenomena," in Proceedings of the the SPE Eastern Regional Meeting, pp. 21–23, Charleston, WV, USA, October 2014, paper SPE 171008. ·
8. M. O. Eshkalak, U. Aybar, and K. Sepehrnoori, "An economic evaluation on the re-fracturing treatment of the U.S. shale gas resources," in Proceedings of the SPE Eastern Regional Meeting, Charleston, WV, USA, October 2014, paper SPE 171009. ·
9. M. O. Eshkalak, S. D. Mohaghegh, and S. Esmaili, "Synthetic, geomechanical logs for marcellus shale," inProceedings of the SPE Digital Energy Conference and Exhibition, The Woodlands, Texas, USA, March 2013, paper SPE 163690.
10. M. Omidvar Eshkalak, Synthetic geomechanical logs and distributions for marcellus shale [M.S. thesis], West Virginia University Libraries, West Virginia University, Morgantown, WV, USA, 2013.

11. L. Rolon, S. D. Mohaghegh, S. Ameri, R. Gaskari, and B. McDaniel, "Using artificial neural networks to generate synthetic well logs," Journal of Natural Gas Science and Engineering, vol. 1, no. 4-5, pp. 118–133, 2009. · ·
12. S. Mohaghegh, R. Arefi, S. Ameri, K. Aminiand, and R. Nutter, "Reservoir characterization with the aid of artificial neural network," Journal of Petroleum Science and Engineering, vol. 16, pp. 263–274, 1996.
13. S. Mohaghegh, M. Richardson, and S. Ameri, "Virtual magnetic imaging logs: generation of synthetic MRI logs from conventional well logs," in Proceedings of the SPE Eastern Regional Meeting, pp. 223–232, Pittsburgh, Pa, USA, November 1998.
14. I. A. Basheer and Y. M. Najjar, "A neural network for soil compaction," in Proceedings of the 5th International Symposium on Numerical Models in Geomechanics, G. N. Pande and S. Pietrusczczak, Eds., pp. 435–440, Balkema, Roterdam, The Netherlands, 1995.
15. A. J. Maren, C. T. Harston, and R. M. Pap, Handbook of Neural Computation Applications, Academic Press, San Diego, Calif, USA, 1990.
16. Y. Khazani and S. D. Mohaghegh, "Intelligent production modeling using full-field pattern recognition,"SPE Reservoir Evaluation & Engineering, vol. 14, no. 6, pp. 735–749, 2011.

Chapter 3

US Natural Gas Market Classification Using Pooled Regression

Vyacheslav V. Kalashnikov[1,2,3], Gerardo A. Pérez-Valdés[4], Timothy I. Matis[5], and Nataliya I. Kalashnykova[3,6]

[1]Departamento de Ingeniería Industrial y de Sistemas, Tecnológico de Monterrey, Avenida Eugenio Garza Sada 2501 Sur, 64849 Monterrey, NL, Mexico

[2]Department of Experimental Economics, Central Economics and Mathematics Institute (CEMI), The Russian Academy of Sciences (RAS), Nakhimovsky Pr. 17, Moscow 117418, Russia

[3]Department of Electronics and Computing, Sumy State University, Rimsky-Korsakov Street 2, Sumy 40007, Ukraine

information. Following the examples found in the literature and our own experience, we revealed that including more explicative series provided very good results in terms of regression fit. This has led to the model presented in the next section.

Coming back to the partitioning method, the two approaches presented before were as follows.

- The first one is the Dendrogram Grouping Method, which "cuts" a binary tree (whose nodes represent regression parameters) based on how close to each other the parameters are with respect to a given metric function and a weight scheme for the entries. This method proved replicative and fast, but it does not provide statistical significance to the grouped states' parameters (i.e., one state might find that temperature is a significant regressor, whereas some other state in the same group may not).
- Another one is a greedy heuristic that starts with a number of states called "group leaders," and iteratively selects for each remaining state the group that suits the state best, based on its regression coefficient R^2 Because of the large amount of regressions performed, this method was reported to be slower and subject to accidental fluctuations, but the final results always guaranteed that all states in one group shared the same significance in their parameters.

In the following sections, we explain how we have improved over our latest approach, adding explicative power and robustness to the partitioning method and, ultimately, creating a better technique to identify similar regions based on their econometric data.

REGRESSION ANALYSIS

Individual Multiple Linear Regression (IMLR)

Let n be the total number of states, m the number of observations per time series (months, in this case), I={1,2,...,n} the set of the 48 contiguous states of the American Union, $t \in T=\{1,2,\ldots,m\}$ the (discrete) time parameter, $\{P_{i,t}\}$ the differenced residential natural gas price in state $i \in I$ at time $t \in T$,$\{T_{i,t}\}$ the differenced temperature, in Kelvin,

Chapter 3

US Natural Gas Market Classification Using Pooled Regression

Vyacheslav V. Kalashnikov[1, 2, 3], Gerardo A. Pérez-Valdés[4], Timothy I. Matis[5], and Nataliya I. Kalashnykova[3, 6]

[1]Departamento de Ingeniería Industrial y de Sistemas, Tecnológico de Monterrey, Avenida Eugenio Garza Sada 2501 Sur, 64849 Monterrey, NL, Mexico

[2]Department of Experimental Economics, Central Economics and Mathematics Institute (CEMI), The Russian Academy of Sciences (RAS), Nakhimovsky Pr. 17, Moscow 117418, Russia

[3]Department of Electronics and Computing, Sumy State University, Rimsky-Korsakov Street 2, Sumy 40007, Ukraine

[4]Institutt for Industriell Økonomi og Teknologiledelse, Norges Teknisk Naturvitenskapelige Universitet, Alfred Getz Vie 3, 7491 Trondheim, Norway

[5]Department of Industrial Engineering, Texas Tech University (TTU), Lubbock, TX 79409, USA

[6]Departamento de Ciencias Físico-Matemáticas (FICA), Universidad Autónoma de Nuevo León (UANL), Avenida Universidad S/N, 66450 San Nicolás de los Garza, NL, Mexico

ABSTRACT

Natural gas marketing has considerably evolved since the early 1990s, when a set of liberalizing rules were passed in both the United States and the European Union that eliminated state-driven regulations in favor of open energy markets. These new rules changed many things in the business of energetics, and therefore new research opportunities arose. Econometric studies about natural gas emerged as an important area of study since natural gas may now be sold and traded in a number of stock markets, each one responding to potentially different behavioral drives. In this work, we present a method to differentiate sets of time series based on a regression model relating price, consumption, supply, and other factors. Our objective is to develop a method to classify different areas, regions, or states into groups or classes that share similar regression parameters. Once obtained, these groups may be used to make assumptions about corresponding natural gas prices in further studies.

INTRODUCTION

In the early 1990s, several regulations were passed in the US and the European Union [1–3] changing the way natural gas was marketed and traded. Particularly, this liberalization [4] effectively ended a period in which natural gas was a state-driven industry. The liberalization has also created the emergent natural gas markets, as well as a strong demand for models to better tackle the new problems and profit from this new setting [5, 6].

Owing not only to this liberalization, but also to the new local conditions that are more open to competition, new small players entered the natural gas industry, especially at the local scale. Indeed, the US has over 80 interstate, long-distance pipelines [7], serving different regions with various climatic, demographic, economic, and political circumstances. Natural gas usage in Alabama, for example, intuitively is not the same as in Oregon; thus the market dynamics of the fuel are also different, and this, we presume, should be reflected in some way in the econometric data of the states.

Not only macroeconomic trends, however, are affected by this setting. When doing cross-regions studies of various aspects of the supply chain, such as the forecasting of demand [8, 9], the balancing of the pipelines after imbalances have been created by the natural gas shippers [10–12], or the dynamics of interstate-intrastate systems [13], one has to take into account the existence of different markets. The existence of a common relationship between price and consumption of natural gas across several zones allows for strong claims of uniformity, which are useful when, for example, we are building scenarios for a stochastic problem. Indeed, if we manage to group the regions in clusters with similar price and consumption functions, we can reduce the number of variables needed in a scenario tree formulation [6, 14].

As such, we specify a regression function that relates many of the most relevant econometric figures for each of the 48 contiguous states of the American Union, modeling price as a function of explicative variables such as natural gas consumption, supply, and storage levels, as well as population (number of costumers), oil prices, temperatures, and production. The regression coefficients are then used to divide the set of states into several subsets or groups, obtaining a partition in which all the states in a group share the same regression parameters, and thus can be classified as an (implicit) market. The partition is made considering both statistical and nonstatistical characteristics of the obtained regression coefficients. The resulting partitions are next compared with others in their similitude and statistical significance, which would validate the goodness of the combination of the dendrogram and GRASP grouping methods.

This paper is organized as follows. The motivation and literature review on natural gas econometric regression is given in Section 2. Section 3 describes the way the regression function is derived, while

Section 4 details the method for using the said function to perform the classification. Section 5 presents and discusses the results of the study, and conclusions are given in Section 6.

NATURAL GAS PRICE-CONSUMPTION MODEL

This work was motivated by our previous research in the natural gas supply chain, specifically developing an optimization model that addresses issues in interstate pipelines. The data used in this model, however, came from different regions, and therefore the time series involved did not necessarily behave in the same way.

As an example, suppose we are trying to model a certain problem that involves forecasting the residential consumption and price of natural gas in the states of Washington and Oregon, that is, four time series. If the robustness of the model is also a concern, then we should additionally consider different forecasting scenarios. Even with only two possible forecasting scenarios for each series (high/low consumption or prices) this translates into 2^4 possible behaviors of the econometric parameters. If consumption is expressed as a function of price, however, then the scenario tree has only 2^2 branches. Furthermore, if the regression function for both states is the same, then the number of scenarios can be reduced to just two. As the number of states being modeled scales up, that is, there are more than two parameters of interest, common assumptions like those mentioned above help reduce greatly the amount of scenarios in a stochastic model, optimization, or otherwise.

As we studied particular sets of data, it was noted that historical data of consumption and price showed conspicuous properties that could be used for the sake of our aims. Even though these data collections were taken from different states, all pairs of time series showed elastic consumption/demand [15, 16]; exponentially growing price averages [15, 17]; and both series in every pair seemed to be highly correlated to each other.

Indeed, the possibility of characterizing one set of series as a (regression) function of the other was interesting, as it would reduce the amount of data we needed to consider when modeling optimization problems. It is, of course, a common practice in economic and managerial sciences to do that since, for example, demand data is simpler to work with than price data [18]. The latter is mainly because the demand is usually easier to predict, and its behavior is less chaotic than that of prices. Such historical relationships between price and consumption is a common topic of study in time series economic analysis [19], which is mostly performed with the inclusion of other explicative variables, such as the price of substitutes (electricity, coal) and weather conditions.

This is the case of several models where the calculation of elasticities is the primary goal of the study [20]. Log-linear models [21–24] are generally favored because of the ease they provide when computing elasticity figures. However, linear models also have applications in the natural gas industry, like the Short-Term Integrated Forecasting system (STIFS) used by the United States Energy Information Agency in order to estimate natural gas demand as a function of several types of important variables related to the energy industry [25].

Former and Current Approaches

As explained in our previous work [26], a carefully designed regression function can help achieve such strong assumptions. Nevertheless, the study of such relationships and the possibility of forming state clusters based merely upon time series data analysis turned out to be interesting by itself, and we developed two different approaches to partition the collection of states. As we observed, neighboring states showed a large amount of diversity, yet different methods of grouping seemed to place certain states consistently together.

Two major areas of opportunity discovered were the design of the regression function, and the trade-off that each partition algorithm made use of.

Our previous paper [26] aimed at a very definite objective regarding the qualities of the regression model: it had to correlate consumption and price of residential natural gas series, using the former as the explicative variable because of the ease in its forecasting. The expression thus obtained served its purpose well, as demonstrated in its application to the optimization models in [27]; nevertheless, a more inclusive approach would involve series that comprise more information. Following the examples found in the literature and our own experience, we revealed that including more explicative series provided very good results in terms of regression fit. This has led to the model presented in the next section.

Coming back to the partitioning method, the two approaches presented before were as follows.

- The first one is the Dendrogram Grouping Method, which "cuts" a binary tree (whose nodes represent regression parameters) based on how close to each other the parameters are with respect to a given metric function and a weight scheme for the entries. This method proved replicative and fast, but it does not provide statistical significance to the grouped states' parameters (i.e., one state might find that temperature is a significant regressor, whereas some other state in the same group may not).
- Another one is a greedy heuristic that starts with a number of states called "group leaders," and iteratively selects for each remaining state the group that suits the state best, based on its regression coefficient R^2 Because of the large amount of regressions performed, this method was reported to be slower and subject to accidental fluctuations, but the final results always guaranteed that all states in one group shared the same significance in their parameters.

In the following sections, we explain how we have improved over our latest approach, adding explicative power and robustness to the partitioning method and, ultimately, creating a better technique to identify similar regions based on their econometric data.

REGRESSION ANALYSIS

Individual Multiple Linear Regression (IMLR)

Let n be the total number of states, m the number of observations per time series (months, in this case), $I=\{1,2,\ldots,n\}$ the set of the 48 contiguous states of the American Union, $t \in T=\{1,2,\ldots,m\}$ the (discrete) time parameter, $\{P_{i,t}\}$ the differenced residential natural gas price in state $i \in I$ at time $t \in T$,$\{T_{i,t}\}$ the differenced temperature, in Kelvin, shifted so that the minimum figure is e, $\{O_t\}$ the differenced average spot price of oil in the US at time $t \in T$, $\{N_{i,t}\}$ the differenced number of residential consumers of natural gas in state $i \in T$ at time $t \in T$, and $\{C_{i,t}\}$ the differenced consumption of natural gas in state i at time t.

Notice that all these series are differenced, or more precisely, lag- (-1) differenced from the original values. This is because the said original values all tested positive for unit roots in the advanced Dickey-Fuller test. In contrast to the original series, the differenced series prove to be stationary; hence we make use of the latter.

This is the linear model we devised to relate the above-described series:

$$\widehat{C'_{i,t}} = \alpha_{0,i} + \alpha_{1,i}P'_{i,t} + \alpha_{2,i}C'_{i,t-12} + \alpha_{3,i}T'_{i,t} + \alpha_{4,i}O'_t + \alpha_{5,i}N'_{i,t}; \quad t \in \mathbf{T}; \quad i \in I. \tag{1}$$

We choose a Robust Regression Analysis using Huber weights to fit the series over traditional least-squares method due to nonnormality of the residuals experienced with the latter. Furthermore, due to the steps described in the next sections, heteroskedasticity would likely appear in the residuals once the pooling regression is carried on.

While most of the series were reasonably fit by (1), a couple of them showed very erratic behavior in either their natural gas price or consumption series. This is expected insofar economic forecasting is commonly subject to the large instability at time t. As the driving force behind short-term fluctuations in natural gas pricing is consumer demand rather than production supply, price was shown to be a significant factor when describing market consumption.

The selection of the descriptive variables was made considering other consumption models in the literature, the available data, and the significance found in the preliminary regression analysis. In particular, electricity prices and the natural gas supply and production, as well as a time index, were tested but found not to be significant in most of the states. This was especially interesting in the case of electricity prices, which certain sources cite as usual descriptors for the natural gas demand, but which were found to be 0.05 significant in only 12 of the 48 cases and thus dropped from the model.

The consumption and price of natural gas are endogenous variables as both are correlated to system shocks, such as unstable governments or weather-related events. As an alternative to the use of least squares regression to fit the model given in (1), a two-stage least squares approach could be employed with such instrumental variables as the number of gas producing wells, reserve estimates, and underground storage, to name only a few. However, this approach is not considered here, because the response (reaction) time of consumers' consumption habits to the shocks is much longer than that to the spot prices set by the market every day.

Pooled Multiple Linear Regression (PMLR)

Now we address the issue of how one can use the same regression formula for more than one state, which would create several classes of states where demand responds to changes in the descriptors in a similar mode.

Assume that we have split n collections of state time series into several classes, with the members of each class sharing a common set of regression parameters. Then the pooled data from the groups would be regressed at the same time, creating pooled regressions.

Let I= $\{I_1, I_2, \ldots, I_k\}$ be a partition of I, and consider the model:

$$\begin{aligned}\widehat{C'_{i,t}} &= \beta_{0,i} + \beta_{1,k}P'_{i,t} + \beta_{2,k}C'_{i,t-12} + \beta_{3,k}T'_{i,t} \\ &\quad + \beta_{4,k}O'_t + \beta_{5,k}N'_{i,t}; \quad t \in T, \\ &\quad \forall i \in I_k, \quad k = 1, 2, \ldots, K. \end{aligned} \tag{2}$$

Note that this model—called the Pooled Multiple Linear Regression (PMLR) model—has K.sets of parameters for each regressor variable, except for the intercepts a_0^i, which we allow to be different for each state. In comparison, model (1) has n sets of parameters.

How should one define the partition I of the set of states? A good partition is expected to deliver groups of more or less congruent sizes, while maintaining a high individual R^2 value for each state. A good partition method should also be replicative (i.e., the same partition is obtained for the same group of states), be fast enough, and support the statistical significance.

DENDROGRAM-GRASP GROUPING METHOD (DGGM)

In this section, a combination of both grouping methods mentioned in [26] into a GRASP heuristic is proposed. The resulting technique inherits the replicative property of the dendrogram method, while retaining the statistical significance of the heuristic algorithm.

Dendrograms

Dendrograms. Dendrograms are binary trees in which two observation vectors a and b form the (sub-)branches of a higher branch c, so that

- these two observation vectors are "closer" to each other than to any other observation d,
- c is not an observation *per se*, but a new, artificial vector formed by some linear combination of a and b.

The term "closer" is interpreted with respect to some metric (e.g., the Euclidean metric), while the artificial observations are produced by the weighted combination method. Once the dendrogram is formed, it is cut down from the root and thus generating (sub-) dendrograms with the branches resulting from the cut. The height of the cut is determined according to one of several criteria (the number of subdendrograms produced, the maximum allowed membership for the subdendrogram, etc.) The leaves pertaining to a given subdendrogram will pool their regression data together and form one group for the PMLR.

Previous experiments [26] have shown that what is called the "average euclidean" metric [28] delivers satisfactorily high R^2 levels with a better homogeneity in the resulting groups than other linkage function options.

GRASP Heuristics

GRASP stands for Greedy Randomized Adaptive Search Process; it is a metaheuristic, that is, a general method designed to provide good—but not necessarily optimal—results in problems otherwise too complicated to find an optimal solution, especially combinatorial problems [29].

Summarily, our GRASP approach will start with a seed formed by several one-state groups; then, for each state, it will identify those groups that deliver higher R^2 values once the data for the current state is pooled with that of the group. This is called the Restricted List of Candidates or RLC. A group I_k from the RLC is chosen at random, and the current state is added to I_k, pooling its data with those already in the group. A number of swaps and movements are performed once the states are all in place, in order to try to improve the values of the resulting statistics R^2.

It is important to note that setting the values for the GRASP routine is rather subjective, since there is no definite objective to be achieved. Indeed, one cannot determine what number of groups is optimal, or which way is the best to define the greedy function. For example, one could prefer to increase the grouped R^2 value in each group rather than the average of the individual R^2s in that group, or vice versa.This is exemplified by the function

$$F_w(I_k) = \omega R^2_{I_k} + \frac{1-\omega}{|I_k|}\sum_{i\in I_k} R^2_i, \tag{3}$$

Where

$$R^2_{I_k} = 1 - \frac{\sum_{t\in \mathrm{T}, i\in I_k}(y_{i,t}-\hat{y}_{i,t})^2}{\sum_{i\in \mathrm{T}, i\in I_k}(y_{i,t}-\overline{y}_{I_k})},$$

$$R^2_i = 1 - \frac{\sum_{t\in \mathrm{T}}(y_{i,t}-\hat{y}_{i,t})^2}{\sum_{t\in \mathrm{T}}(y_{i,t}-\overline{y}_i)}. \tag{4}$$

Here, $y_{i,t} = \ln C_{i,t}$, and $\bar{y}_\iota$ is understood as the average of all of the observations belonging to ι if the latter is a state (e.g., $\iota = i$) or as the average of the observations of the states in ι, if the latter is a set of states (e.g., $\iota = I_k$). For the local search, we handle the improvement function $G\tau(I_k, I_\ell, I_i)$, which is used when deciding if it is convenient to move state i from group k to group ℓ. It is parametrized by the improvement weight τ:

$$G_\tau(I_k, I_l, I_i) = (1-\tau)\frac{R^2_{I_k} + R^2_{I_\ell}}{2} + \tau R^2_i. \quad (5)$$

Dendrogram-GRASP Algorithm

The following algorithm is used to classify the set of 48 contiguous states of the United States into groups that share a common regression function.

- Initialize the values for each of the time series in each of the 48 states. Set a seed size s_{Seed}, a maximum number of groups s_{Max}, a RLC size s_{RLC}, an individual/grouped R^2 weight $\omega \in [0,1]$, an individual/ grouped threshold $\varphi \in (0, 1)$, an improvement weight $\tau \in (0, 1)$, a relative improvement threshold $\psi \in [0,1]$, and a maximum number of local search steps, s_{ls}.

Seed Selection

- Performan IMLR on each of the 48 sets of time series, obtaining $\alpha_{j,i}$, $j = 0, \ldots, 6$, $i \in I$.
- Form a dendrogram of 48 leaves with the vectors α, using the average euclidean mean as the linkage function, and cut it so that there are exactly s_{Seed} subtrees.
- Select the state with the highest R^2_i from each of the obtained groups and call it the kth group's leader. Define the one-state groups obtained as the partition $\mathbf{I}_k$. All the nonselected (spare) states form the set *ActiVe*.

Greedy Process

- For each state x in the set $Active$,
    1. Pool the data of x with the data of each of the formed groups and perform a pooled Mathematical Problems in Engineering 5 regression; select a number of s_{RLC} groups with the highest value of the greedy function F_w and form the RLC;
    2. Choose randomly one of the groups from the RLC, for example, I_a.
        a. If none of the candidate groups in the RLC delivers $F_w() > \varphi$ and we have not yet reached the maximum number of groups s_{Max}, create a new group $I_x = \{x\}$ containing only x, remove x fromthe active set, and update all the parameters.
        b. Otherwise, assign x to I_a, remove x from the active set, and update all the parameters.
- All of the states are now partitioned into the groups, and we can begin the local search.

Local Search

- For $l = i$ to $l = s_{ls}$,
- randomly select one of the formed groups, I_a, and one state in that group, x; select another group, I_b; compute $g_1 = G_\tau (I_a, I_b, x)$;
- remove x's data from $I\ a$ and pool the same data of x with I_b; compute $g_2 = G_\tau (I_a, I_b, x)$;
- if $g_1 \geq (1 + \psi) g_2$, remove x from $I\ b$ and return it to I_a; otherwise, continue.
- Report the obtained groups as the desired partition.
- End.

Partition Similarity

To determine the similitude of two partitions, we will use an expression that, roughly speaking, counts the number of coincidences found in two partitions and divides it by the number of total possible coincidences, given the sizes of the groups in each partition. While there are many disputable ways to measure the similitude between partitions with a different number of elements, this method was chosen because of its normality. Indeed, it will always return 1 when both partitions are identical and will always return 0 when there are no coincidences between two partitions, that is, when no two states share a group in both partitions and no state is single-grouped in both partitions.

Let $\mathbf{I} = \{I_1, I_2, \ldots, I_K\}$, $\mathbf{J} = \{J_1, J_2, \ldots, J_L\}$ be two arbitrary partitions of the set of states, with $I_i = \{I^i_1, I^i_2, \ldots, I^i_{k^i}\}$, $i = 1, \ldots, K$, and $J_j = \{J^j_1, J^j_2, \ldots, J^j_{l^j}\}$, $j = 1, \ldots, L$. The function $a_{I,J}$ defined by

$$\mathbf{a}_{\mathbf{I},\mathbf{J}}(I_i) = \begin{cases} 1, & \text{if } I_i = \{m\} = J \text{ for any } J \in \mathbf{J}, \\ 0, & \text{otherwise,} \end{cases} \qquad I_i \in \mathbf{I}, \tag{6}$$

assumes the value 1 if group I_i contains a single state in partition **I** and this state also forms a group-singleton in partition J.

For every pair of states, wewill assess if they share a group in a given partition using the following function $\mathbf{b}_{\mathrm{J}}$:

$$\mathbf{b}_{\mathbf{J}}(m,n) = \begin{cases} 1, & \text{if } m,n \in J_j, \text{ for any } j; \\ 0, & \text{otherwise,} \end{cases} \qquad m,n \in \mathbf{I}. \tag{7}$$

In case the function $a_{I,J}$ has the value of 1, we say that we have a (one-state) coincidence, which means that the state has been found incompatible with other states twice, no matter which method formed partitions I, J.

Similarly, if the function b_J returns 1 for two states *in a group from the partition* **I**, we say that we have a (two state) coincidence; that is, in both partitions, the two states are members of the same group.

To measure the number of coincidences between two partitions, we use the function:

$$C_q(I_i, \mathbf{I}, \mathbf{J}) = \mathbf{a}_{\mathbf{I},\mathbf{J}}(I_i) + (1 - \mathbf{a}_{\mathbf{I},\mathbf{J}}(I_i)) \times \left(\sum_{m \in I_i} \sum_{n \in I_i, n \neq m} \frac{q\mathbf{b}_{\mathbf{J}}(m,n) + (1-q)}{2} \right), \tag{8}$$

for $I_i \in \mathbf{I}$, $q = \{0,1\}$.

If the parameter q equals 1, then the function C_q counts the number of either type of coincidences that couples of states reveal in the group I_i in comparison to the groups they belong to in the partition **J**. Conversely, if $q = 0$, then we simply count the total number of possible coincidences for the states in the group $I_i \in \mathbf{I}$. Note that the function C_q is not necessarily symmetric with respect to the pairs of partitions: $C_q(I_i, \mathbf{I}, \mathbf{J})$ need not have the same value as $C_q(Ii, \mathbf{J}, \mathbf{I})$.

The similitude function used Sim is defined as follows:

$$\mathrm{Sim}(\mathbf{I}, \mathbf{J}) = \frac{\sum_{I_i \in \mathbf{I}} C_1(I_i, \mathbf{I}, \mathbf{J}) + \sum_{J_j \in \mathbf{J}} C_1(J_j, \mathbf{J}, \mathbf{I})}{\sum_{I_i \in \mathbf{I}} C_0(I_i, \mathbf{I}, \mathbf{J}) + \sum_{J_j \in \mathbf{J}} C_0(J_j, \mathbf{J}, \mathbf{I})}. \tag{9}$$

Notice that if there is at least one group in either partition containing more than one element, then C_0 for that group is at least 1, whereas if there exists no such group in either partition, then $a_{I,J}(a) = 1$ and consequently $C_0(a,I,J) = 1$ for any $a \in I \cap J$. Therefore, the denominator is never 0, which makes this function well defined.

Lemma 1: *Let* I *and* J *be two partitions of the set* $I = \{1, 2.....,\}$, *and let function* Sim *be defined by* (9). *The following statements are true:*

- Sim (I, J) = Sim (J, I);
- Sim (I, J) = 1 *if and only if* I = J;
- Sim (I, J) = 0 *if and only if there are neither one-state nor two-state coincidences between* I *and* J;
- 0 ≤ Sim (I, J) ≤ 1.
- *Proof.* (1) This is easy to see from the structure of the function.

- Let I = J.If $I_i = \{m\} = J_k$ for some i and k, then C_1 (I_i,I_j) = C_0 (I_i, I,J).Otherwise, if the order of I_i is greater than one, then the second term in (8) (the definition of C_q) assumes the same value no matter whether $q = 1$ or $q = 0$. Therefore, the numerator and denominator in Sim are equal. Conversely, if there exists one $I\ i$ such that $I_i \neq J$ for all $J \in$ J, then C_1 (I_i, I, J) is strictly less than C^0 (I, I, $_J$). Since C_1 ($J_{i'}$, I, J) ≤ C_0 (I_i, I, J), it follows that the numerator in (9) (defining Sim) is strictly smaller than the denominator, and therefore Sim(I, J) < 1.
- If there is at least one one-state coincidence, or a twostate coincidence, then the numerator in Sim is larger than 0, and therefore Sim(I, J) > 0.
- Conversely, since C_q is nonnegative for every value of q, Sim(I, J) = 0 means that both terms in the numerator are zero, which is only possible if $a_{I,J}(I_i) = a_{I,J}(J\ j) = 0$ for every member of I and J, *and* bJ(m, n) = 0 for every m, $n \in I$, which means that there is no coincidence of any type between these two partitions.
- The first inequality follows fromthe fact that both the numerator and denominator in (9) are positive. The second inequality comes from the same argument as in item (2); that is, the numerator is either equal or strictly less than the denominator.

EXPERIMENTAL RESULTS

This section presents the results of the numerical experimentation performed on a number of times series pertaining to each of the 48 data sets. The values for the historical natural gas prices, consumption, and number of consumers, as well as the oil spot prices were taken from the US Energy Information Agency, whereas the temperature figures for each state were obtained from the US Department of Commerce National Oceanographic and Atmospheric Agency [30].

IMLR Results

The first step was to performthe IMLR for the 48 sets of time series; this provided the regression parameters for the dendrogram formation. The five time series corresponding to every state had 240 monthly observations each.

Individual regression models showed regression R^2 coefficients with the average of 0.77 and theminimumof 0.61.The normality and heteroskedasticity were not tested due to the use of Robust Regression with Huber weights. Randomness of the residuals was tested, and high *P* values were found for many states.

Dendrogram-GRASP Grouping Results.

There are two main aspects we wanted to consider when evaluating the effectiveness of the Dendrogram-GRASP approach: how replicative it is, and how good a partition is produced. The first issue is evaluated by examining how good and how similar the partitions are that come from the same seed (as opposed to those that come fromrandomly generated seeds). The goodness of one partition is measured with the average group [state] coefficient of determination, $R^2_{I_k}$ [R^2_I], calculated across all the groups [states] of the partitions.

There are, however, a number of different design parameters that should be included in the experimentation. Each experimental observation consists of the generation of 10 partitions, using the following parameters.

- Aseed choice: the dendrogramseed (DDR), a random seed common to all 20 partitions (FIX), and a random seed for each partition (RND).
- The individual versus grouped R^2 weight, ω, which determines what is more important when adding a new state to an existing group in the GRASP routine: values considered in the experimentation are $\omega = 0$ (only the single states' R^2s are considered), 0.5, and 1 (only the groups' R^2s are important).
- The new group threshold, φ: the closer the value of φ to 1, the more likely new single-state groups will be created in the GRASP routine. The tested values are $\varphi \in \{0.90, 0.95\}$.
- The length of the restricted candidate list, s_{RCL}: the values considered are $s_{RLC} \in \{1, 5\}$.
- The number of local search moves: $s_{ls} \in \{0, 100\}$.
- (The local search individual/grouped R^2 weight, τ: considered values are $\tau \in \{0, 0.66, 1\}$.

The starting number of groups was fixed at 10, and the maximum number of groups allowed was set at 15. Each combination of levels was replicated 20 times. This resulted in 5760 experimental observations.

In each observation, we calculated the average similitude between the various partitions involved, as well as their similitude with a randomly created partition. The compared similitudes were as follows:

- the average similitude of the dendrogram partition to each of the 20 GRASP partitions (DG);
- the average similitude of a random partition and each of the 20 GRASP partitions (GR);
- The average similitude of the 20 GRASP partitions among themselves (GG).

The first part of the analysis consisted in testing all the experimental observations. After that, only the most convenient levels were kept.

Tables 1 and 2 present a summary of the results of the experimental runs. The first three data columns show the average similarities for each of the three comparisons of interest, whereas the last two columns show the average of the individual and grouped coefficients of determination.

Table 1: Experimental Results I

Factor	Level	Av. similitude			Av. R2 values	
		DG	GR	GG	Av. $R^2_{I_i}$	Av. $R^2_{I_k}$
φ	0.90	0.145	0.079	0.178	0.503	0.535
	0.95	0.149	0.079	0.177	0.499	0.537
Seed	DDR	0.182	0.083	0.194	0.513	0.568
	FIX	0.130	0.077	0.154	0.489	0.521
	RND	0.128	0.077	0.184	0.501	0.520
ω	0	0.146	0.082	0.136	0.427	0.564
	0.5	0.147	0.079	0.183	0.534	0.535
	1	0.148	0.077	0.213	0.542	0.511
τ	0	0.160	0.080	0.232	0.699	0.502
	0.66	0.141	0.079	0.150	0.455	0.554
	1	0.140	0.079	0.149	0.349	0.553

Table 2: Experimental Results II

Factor	Level	Av. similitude			Av. R2 values	
		DG	GR	GG	Max. $R^2_{I_i}$	Max. $R^2_{I_k}$
SRLC	1	0.160	0.082	0.247	0.879	0.862
	5	0.134	0.076	0.107	0.879	0.871
S1s	0	0.167	0.084	0.236	0.876	0.857
	100	0.126	0.075	0.119	0.882	0.876

A quick look at this table suggests that the similitude figures are characteristically low: the average similarity of an arbitrary partition to a randomly formed one, calculated using all the observations, is 0.0947. This will be called the partitions' randomness. If columns 3 and (particularly) 5 approach the average randomness for this experiment, the partition method is not very efficient. This especially concerns the cases $s_{ls} = 5$, $\omega = 0$, and $\tau = 1$, whose similarity measures are fairly low. Luckily enough, in all these cases the averageGGsimilarities were found to be statistically different (higher) than their respective GR similarities by making use of the Wilcoxon signed-rank (WSR) $\alpha = 0.95$ test.

The average R^2 values in columns 6 and 7 do not deviate much from the averages across all the observations, 0.602 and 0.624, respectively, with the exception of the grouped individual parameter R^2_{Ii} for $\tau = 1$. It is clear that certain similarity values for some levels are consistently lower than others.There is, for example, a very large difference between the average DG similitude obtained using a DDR seed than using aRNDor FIX seed and so on. Based on this,we decided to discard some of the levels whose averages are not only considerably lower, but also the observations for each level are determined to be different by aWSR test.

Now let us look at each of the level values we should consider to drop. The first level, the GRASP new group threshold φ, shows a very similar GG figure, and equally similar R^2 values. We decide to keep the factor levels intact, in case these figures change once other levels are removed.

Seeds are more difficult to assess. The FIX seed shows lower values than theDDGone, but still higher than theRND. Weight τ shows much better numbers in all but the grouped R^2 entry. Because of this, we pick it as the only label for the later study.On the contrary, ω is better at value 1, except again in the grouped R^2 column. This result for ω is very counter intuitive! However, the two values serve a similar purpose at different parts of the process, so this behavior might indeed be justified.

The factors s_{RLC} and s_{ls} were introduced to add variation in the GRASP routine, and their results appear separated in Table 2. This is because, while their similitude values work in the same way as the other factors, the R^2 measurements per observation are not the average across all 10 partitions in the observation, but rather the maximum obtained. In a common GRASP routine, the process will be repeated several times and the best solution will be adopted. For our case, this means that we should choose the best of the 20 partitions in each observation, and this decision becomes the result for that observation. Arguably, both the individual and grouped average maximum coefficients of determination seem to show little difference. In particular, the differences are deemed not large enough to justify the trade-off with similarity in all cases. While this was expected from the extended RLC size, the poor results obtained by the local search suggest that we should rethink our local search procedure in the future.

Based on similarity alone, we decided to eliminate the poorest levels and kept only a single-group state list and a zero-swaps local search for the second part of the analysis. After deciding to drop several levels, we will rewrite the results table including only the accepted levels, to see how the figures change once the poorest results are winnowed.

The much smaller Table 3 is the consequence of fixing ω=1, τ=0, s_{RLC}=1,and S_{ls}=0and eliminating the RND seed choice, which results in 100 observations. Now the similitudes look much better: we have the sample average of 0.438 and the maximum of 0.477, which means that, for the parameters chosen, the similitudes obtained are remarkably higher than the average randomness.

Table 3: Experimental Results III

Factor	Level	Av. similitude			Av. R2 values	
		DG	GR	GG	Av. R_i^2	Av. R_{ik}^2
	0.90	0.171	0.077	0.432	0.760	0.340
	0.95	0.178	0.085	0.432	0.759	0.349
Seed	DDR	0.238	0.090	0.454	0.757	0.432
	FIX	0.143	0.074	0.365	0.760	0.299
	RND	0.143	0.079	0.477	0.761	0.302

For the first factor, φ, the similitudes are of little difference, the same as the determination coefficients in all accounts. However, for the seed levels, the DDR seed clearly favors similitude between the seed and the resulting partition. Similitude among resulting partitions is also good at theRND partition, which could indicate the particular FIX seed was initially a bad choice when compared to either an average partition seed or one selected in a methodical way.

The coefficients of determination R^2 present a rather interesting development. The individual coefficients R^2_i are decent enough when compared to the ones fromthe dropped levels, but there is a dramatic drop in the group figuresR_{ik}^2, which decreased from an average of around 0.53 to as low as 0.299. This happens because, while focusing on similitude, we chose in favor of $s_{ls} = 0$, which yields the mean R_{ik}^2 of only 0.366, as opposed to the 0.706 value obtained after fixing $s_{ls} = 100$. In Table 2, however, we see the greater max R_{ik}^2 because it was relevant to that table. If we were to remake Table 3 using the value of $s_{ls} = 100$ for this level, similitudes would fall around 10%, but the average group determination coefficients R_{ik}^2 would increase to roughly 0.43, which is much better than that with $s_{ls} = 0$. Maximum values for the different R^2s, correspondent to those in Table 3, remainmostly unchanged.

CONCLUDING REMARKS

In this paper, we propose and justify a heuristic method to group several zones based on a regression function that estimates several factors related to the natural gas demand. The groups thus obtained share key information regarding the behavior of natural gas-related historic econometric data.

We start by developing a linear regression model that correlates natural gas historic residential consumption and several explicative variables, such as the residential price, number of consumers, and temperature. This model, inspired by several examples in the literature, fits well the time series employed and has good predictive power, but it is by no means the only one that can be used nor necessarily the best.

The results of each of the 48 regressions performed are then used to create dendrogram-based partitions, which are in turn used as the starting point in a GRASP routine. The latter, while tending to form rather dissimilar partitions (compared to the dendrogram grouping), has the advantage of adding statistical significance to all the regressions in all the groups formed.

We tested several parameters in an experimental design consisting of more than 4300 observations, six factors, and two or three levels per factor. Using ad hoc and nonparametric selections, we tried to obtain a good combination of parameters, namely, one that delivers high similitude between partitions obtained from the same seed and a satisfactory goodness of the pooled regressions.

Similitude is measured by a standardized function which equals 0 if there are no common groups between two partitions of a fixed set and 1 if both partitions are identical. We were able to obtain experimental conditions with similitudes (mostly) above 0.43, which are deemed good considering that the average randomness of a partition in the study is around 0.09.

It is encouraging that, using the regression function herein proposed, the GRASP routine worked well by itself and also when combined with the dendrogram partitioning method. Unfortunately, the inclusion of randomness did not provide for good results, as it offered no increase in goodness of the partitions but a considerable decrease in similitude when a long RLC was used. The proposed local search approach was found to have a negative impact on the similitude values, though not overly so. However, at the same time it did affect heavily the values of the grouped coefficients of determination when the maximum values were considered in the selection but the averaged values were looked into in the end results. The "goodness" of the regressions, as discussed, must then be judged with a more nuanced approach

The entire work frame summarized here is intended to provide a way to identify individuals (states, in this case) with common econometric behavior among themselves by means of statistically significant information. Such results used to help us in the past in the context of optimization theory (by greatly decreasing the number of variables in stochastic problems), and we believe this technique has other applications in economic analysis.

The planned future work includes enhancing the robustness of the method by designing better GRASP RLC and local search procedures, trying sampled regressions when forming large groups to gain on time and studying how different data sets and regression models would work in combination with the Dendrogram-GRASP approach proposed in the paper.

ACKNOWLEDGMENTS

The research activity of the first author was financially supported by the R&D Department (Cátedra de Investigación) CAT-174 of the Tecnológico de Monterrey (ITESM), Campus Monterrey and by the SEP-CONACYT Project CB-2008-01-106664, Mexico. Also, the work of the fourth author was supported by the National Council of Science and Technology (CONACyT) of Mexico as part of the Project CB-2011-01-169765, PROMEP 103.5/11/4330, and PAICYT 464-10.

REFERENCES

1. Energy Information Administration, "FERC Order 636: The Restructuring Rule," 2005,http://www.eia.doe.gov/emeu/steo/pub/document/textng.html.
2. Energy Information Administration, "FERC Policy on System Ownership Since 1992," 2005,http://www.eia.doe.gov/oil_gas/natural_gas/analysis_publications/ngmajorleg/fercpolicy.html.
3. A. Soto, "FERC Order 636 & 637," 2008,http://www.aga.org/Legislative/issuesummaries/FERCOrder636637.html.
4. IHS Engineering, "EC Proposes New Legislation for European Energy Policy," 2007,http://engineers.ihs.com/news/eu-en-energy-policy-9-07.html.

5. Environmental Protection Agency, "The Impacts of FERC Order 636 on coal mine gas project development," 2008, http://www.epa.gov/cmop/docs/pol004.pdf.
6. K. T. Midthun, Optimization Models for Liberalized Natural Gas Markets, Norwegian University of Science and Technology. Faculty of Social Science and Technology Management Department of Industrial Economics and Technology Management, Trondheim, Norway, 2009.
7. M. J. Doane and D. F. Spulber, "Open access and the evolution of the US spot market for natural gas," Journal of Law and Economics, vol. 34, no. 2, pp. 447–517, 1994.
8. R. Gutiérrez, A. Nafidi, and R. G. Sánchez, "Forecasting total natural-gas consumption in Spain by using the stochastic Gompertz innovation diffusion model," Applied Energy, vol. 80, no. 2, pp. 115–124, 2005.
9. F. K. Lyness, "Gas demand forecasting," Journal of the Royal Statistical Society—Series D, vol. 33, no. 1, pp. 9–12, 1984.
10. S. Dempe, V. Kalashnikov, and R. Z. Ríos-Mercado, "Discrete bilevel programming: application to a natural gas cash-out problem," European Journal of Operational Research, vol. 166, no. 2, pp. 469–488, 2005.
11. V. V. Kalashnikov and R. Z. Ríos-Mercado, "A natural gas cash-out problem: a bilevel programming framework and a penalty function method," Optimization and Engineering, vol. 7, no. 4, pp. 403–420, 2006.
12. N. Keyaerts, L. Meeus, and W. D'haeseleer, "Analysis of balancing-system design and contracting behavior in the natural gas markets," in Proceedings of the European Doctoral Seminar on Natural Gas Research, Delft, The Netherlands, 2008.
13. H. G. Huntington, "Federal price regulation and the supply of natural gas in a segmented field market," Land Economics, vol. 54, no. 3, pp. 337–347, 1978.
14. A. Tomasgard, F. Romo, M. Fodstad, and K. T. Midthun, "Optimization models for the natural gas value chain," in Geometric Modeling, Numerical Simulation, and Optimization: Applied Mathematics at SINTEF, optimization models for the natural gas value chain, Springer, 2007.

15. C. Nelder, "Natural Gas Price ForecastThe future of Natural gas: It's time to invest," 2009, http://www.energyandcapital.com/articles/natural-gas-price-forecast/916.
16. EuroGas, "EuroGas long term outlook to 2030," 2009,http://www.eurogas.org/uploads/media/Statistics_Eurogas_LT_Outlook_2007-2030_Final_25.11.10.pdf.
17. P. W. MacAvoy, The Natural Gas Market: Sixty Years of Regulation and Deregulation, Yale University Press, New Haven, Conn, USA, 2000.
18. K. T. Talluri and G. J. van Ryzin, The Theory and Practice of Revenue Management, Springer, New York, NY, USA, 2004.
19. P. W. Keat and P. K. Y. Young, Managerial Economics: Economic Tools for Today's Decision Makers, Prentice Hall, Englewood Cliffs, NJ, USA, 2006.
20. J. M. Gowdy, "Industrial demand for natural gas. Inter-industry variation in New York state," Energy Economics, vol. 5, no. 3, pp. 171–177, 1983.
21. J. G. Beierlein, J. W. Dunn, and J. G. McConnon Jr., "The demand for electricity and natural gas in the northeastern United States," The Review of Economics and Statistics, vol. 63, no. 3, pp. 403–408, 1981.
22. W. T. Lin, Y. H. Chen, and R. Chatov, "The demand for natural gas, electricity and heating oil in the United States," Resources and Energy, vol. 9, no. 3, pp. 233–258, 1987.
23. N. Krichene, "World crude oil and natural gas: a demand and supply model," Energy Economics, vol. 24, no. 6, pp. 557–576, 2002.
24. S.-H. Yoo, H.-J. Lim, and S.-J. Kwak, "Estimating the residential demand function for natural gas in Seoul with correction for sample selection bias," Applied Energy, vol. 86, no. 4, pp. 460–465, 2009.
25. Energy Information Administration, "Natural Gas Model Description," 1999,http://www.eia.doe.gov/oil_gas/natural_gas/analysis_publications/ngmajorleg/ferc636.html.
26. V. V. Kalashnikov, T. I. Matis, and G. A. Pérez-Valdés, "Time series analysis applied to construct US natural gas price functions for groups of states," Energy Economics, vol. 32, no. 4, pp. 887–900, 2010.

27. V. V. Kalashnikov, G. A. Pérez-Valdés, A. Tomasgard, and N. I. Kalashnykova, "Natural gas cash-out problem: bilevel stochastic optimization approach," European Journal of Operational Research, vol. 206, no. 1, pp. 18–33, 2010.
28. Mathworks Inc, "Statistics Toolbox: Linkage," 2008, http://www.mathworks.com.
29. P. Festa and M. G. C. Resende, "GRASP: an annotated bibliography," in Essays and Surveys in Metaheuristics, pp. 325–367, Citeseer, 2002.
30. "NOAA Satellite and Information Service," 2010 , ftp://ftp.ncdc.noaa.gov/pub/data/cirs/.

Chapter 4

Life Cycle Greenhouse gas Emissions of Marcellus Shale gas

Mohan Jiang[1], W Michael Griffin[2, 3], Chris Hendrickson[1], Paulina Jaramillo[3], Jeanne VanBriesen[1], and Aranya Venkatesh[1]

[1]Civil and Environmental Engineering Department, Carnegie Mellon University, 5000 Forbes Avenue, Pittsburgh, PA 15213, USA

[2]Tepper School of Business, Carnegie Mellon University, 5000 Forbes Avenue, Pittsburgh, PA 15213, USA

[3]Department of Engineering and Public Policy, Carnegie Mellon University, 5000 Forbes Avenue, Pittsburgh, PA 15213, USA

ABTRACT

This study estimates the life cycle greenhouse gas (GHG) emissions from the production of Marcellus shale natural gas and compares its emissions with national average US natural gas emissions produced in the year 2008, prior to any significant Marcellus shale development. We estimate that the development and completion of a typical Marcellus shale well results in roughly 5500 t of carbon dioxide equivalent emissions or about 1.8 g CO_2e/MJ of gas produced, assuming conservative estimates of the production lifetime of a typical well. This represents an 11% increase in GHG emissions relative to average domestic gas (excluding combustion) and a 3% increase relative to the life cycle emissions when combustion is included. The life cycle GHG emissions of Marcellus shale natural gas are estimated to be 63–75 g CO_2e/MJ of gas produced with an average of 68 g CO_2e/MJ of gas produced. Marcellus shale natural gas GHG emissions are comparable to those of imported liquefied natural gas. Natural gas from the Marcellus shale has generally lower life cycle GHG emissions than coal for production of electricity in the absence of any effective carbon capture and storage processes, by 20–50% depending upon plant efficiencies and natural gas emissions variability. There is significant uncertainty in our Marcellus shale GHG emission estimates due to eventual production volumes and variability in flaring, construction and transportation.

INTRODUCTION

Marcellus shale is a rapidly developing new source of US domestic natural gas. The Appalachian Basin Marcellus shale extends from southern New York through the western portion of Pennsylvania and into the eastern half of Ohio and northern West Virginia (Kargbo*et al* 2010). The estimated basin area is between 140 000 and 250 000 km^2 (Kargbo *et al* 2010), and has a depth ranging from 1200 to 2600 m (US DOE 2009). The shale seam's net thickness ranges from 15 to 60 m (US DOE 2009) and is generally thicker from west to east (Hill *et al* 2004). Figure **1** shows the location of the Marcellus and other shale gas formations in the continental United States.

Figure 1: Shale gas plays and basins in the 48 states (source: US Energy Information Administration 2011a.

Shale gas has become an important component of the current US natural gas production mix. In 2009, shale gas was 16% of the 21 trillion cubic feet (Tcf) or 600 million cubic meters (Mm^3) total dry gas produced (US EIA 2011a, 2011b). In 2035, the EIA expects the share to increase to 47% (12 Tcf or 340 Mm^3) of total gas production. The prospect of rapid shale gas development has resulted in interest in expanding natural gas use including increased natural gas fired electricity generation, use as an alternative transportation fuel, and even exporting as liquefied natural gas. To date most shale gas activity has been in the Barnett shale in Texas. However, the immense potential of the Marcellus shale has stimulated increased attention. The shale play has an estimated gas-in-place of 1500 Tcf or 42 000 Mm^3, of which 262–500 Tcf or 7400–14 000 Mm^3 are thought to be recoverable (Hill *et al* 2004, US DOE 2009).

Advancements in horizontal drilling and hydraulic fracturing, demonstrated successfully in the Barnett shale and first applied in the

Marcellus shale in 2004, have enabled the recovery of economical levels of Marcellus shale gas. After vertical drilling reaches the depth of the shale, the shale formation is penetrated horizontally with lateral lengths extending thousands of feet to ensure maximum contact with the gas-bearing seam. Hydraulic fracturing is then used to increase permeability that in turn increases the gas flow.

In this study, life cycle greenhouse gas (GHG) emissions associated with the Marcellus shale gas production are estimated. The difference between GHG emissions of natural gas production from unconventional Marcellus gas wells and average domestic wells is considered to help determine the environmental impacts of the development of shale gas resources. The results of this analysis are compared with life cycle GHG emissions of average domestic natural gas pre-Marcellus and imported liquefied natural gas. In addition domestic coal and Marcellus shale for electricity generation are compared. Other environmental issues may also be of concern in the Marcellus shale development, including disruption of natural habitats, the use of water and creation of wastewater as well as the impacts of truck transport in rural areas. However these environmental issues are outside the scope of our analysis and are not addressed in this paper.

In estimating GHG emissions, we include GHG emissions of carbon dioxide, methane and nitrous oxide. We converted the GHG emissions to carbon dioxide equivalents according to the global warming potential (GWP) factors reported by IPCC. We use the 100-year GWP factor, in which methane has a global warming potential (GWP) 25 times higher than carbon dioxide (IPCC 2007).

MARCELLUS SHALE GAS ANALYSIS BOUNDARIES AND FUNCTIONAL UNIT

The boundary of our analysis and the major process steps included in our estimates are shown in figure 2. Final life cycle emission estimates are reported in grams of carbon dioxide equivalent emissions per megajoule of natural gas (g CO_2e/MJ) produced. Each of the individual processes in the natural gas life cycle has an associated upstream supply chain and is included in this study to provide a full assessment of GHG emissions associated with Marcellus shale gas. The sources of GHG emissions considered in the LCA include: emissions from the production and transportation of material involved in the well development activities (such as trucking water); emissions from fuel consumption for powering the drilling and fracturing equipment; methane leaks and fuel combustion emissions associated with gas production, processing, transmission, distribution, and natural gas combustion.

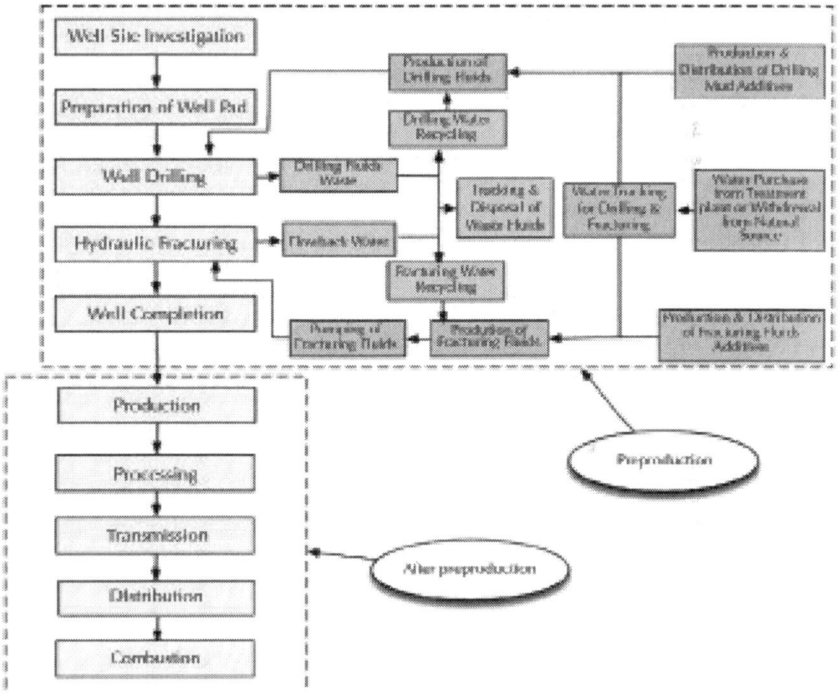

Figure 2: Analysis boundaries and gas production processes.

The life cycle of Marcellus shale natural gas begins with a 'preproduction phase' that includes the well site investigation, preparation of the well pad including grading and construction of the well pad and access roads, drilling, hydraulic fracturing, and well completion (Soeder and Kappel 2009). After this preproduction phase is completed, the well becomes operational and starts producing natural gas. This natural gas can require additional processing to remove water, CO_2 and/or natural gas liquids before it enters the natural gas transmission and distribution system, which delivers it to final end users. For this work we assume that the GHG emissions for production, transmission, distribution and combustion of Marcellus shale natural gas are similar to average domestic gas sources as estimated by Jaramillo *et al* (2007) and further developed and updated by Venkatesh *et al* (2011).

Finally, natural gas has many current and potential uses including electricity generation, chemical feedstock, and as a transportation fuel. Modeling these uses allows comparisons of different primary energy sources. Here we model its use for power generation since it is the largest single use of natural gas in the US (US EIA 2011a, 2011b).

As previously mentioned, this study integrates GHG emissions from the life cycle of water associated with Marcellus shale gas production. Large amounts of water are consumed in the drilling and hydraulic fracturing processes (preproduction phase). Hydraulic fracturing uses fluid pressure to fracture the surrounding shale. The fracturing fluid consists of water mixed with a number of additives necessary to successfully fracture the shale seam. The source of the water varies and can be surface or ground water, purchased from a local public water supplier, or reused fracturing water. In this study we assume 45% of the water is reused on site and the original sources are surface water (50%) and purchased from a local water treatment plant (50%). Regardless of the water source used to produce the hydraulic fracturing fluid, trucks transport the water for impoundment at the well pad. In addition, flowback water (hydraulic fracturing fluid that returns to the surface) and produced water must be trucked to the final disposal site. This water is assumed to be disposed of via deep well injection. A detailed description of the method and data sources used to estimate the GHG emissions associated with all these stages is presented in section 3.

Marcellus shale gas production is in its infancy. Thus, industry practice is evolving and even single well longevity is unknown. Assumptions related to production rates and ultimate recovery have considerable uncertainty. Below, we include a sensitivity analysis for a wide range of inputs parameters.

This study does not consider any GHG emissions outside of the Marcellus shale gas preproduction and production processes. Natural processes or development actions such as hydraulic fracturing might lead to emissions of the shale gas external to a well, particularly in the case of poorly installed well casings (Osborn *et al* 2011). Any such external leaks are not included in this study.

METHODS FOR CALCULATING LIFE CYCLE GREENHOUSE GAS EMISSIONS

Our study used a hybrid combination of process activity emission estimates and economic input–output life cycle assessment estimates to estimate the preproduction GHG emission estimates (Hendrickson *et al* 2006, CMU GDI 2010). Emissions from production, processing and transport were adapted from the literature. We include emissions estimates based on different data sources and reasonable ranges of process parameters. Table 1 summarizes estimation approaches used in this study, while calculation details appear in the supplementary information (available at stacks.iop.org/ERL/6/034014/mmedia).

Table 1: Greenhouse gas estimation approaches and data sources

Process	Estimation approaches	Data sources
Preparation of Well Pad:		
Vegetation clearing	Estimated area cleared multiplied by vegetative carbon storage to obtain carbon loss due to land use change	NY DEC (2009), Tilman *et al* (2006)
Well pad construction	Detailed cost estimate and EIO-LCA model	RSMeans (2005), CMU GDI (2010)
Well drilling:		
Drilling energy consumption	(1) Energy required and emission factor, and (2) cost estimate and EIO-LCA model	Harper (2008), Sheehan *et al* (2000), CMU GDI (2010)
Drilling mud production	(1) Cost estimate and EIO-LCA and (2) emission factors multiplied by quantity.	Shaker (2005), PRé Consultants (2007), CMU GDI (2010)

Drilling water consumption	Trucking emissions plus water treatment emissions multiplied by quantity	Wang and Santini (2009), URS Corporation (2010), PA DEP (2010), Stokes and Horvath (2006)
Hydraulic fracturing:		
Pumping	Pumping energy multiplied by emission factor	URS Corporation (2010), Kargbo *et al* (2010), Currie and Stelle (2010), Sheehan *et al* (2000)
Additives production	Additive quantities cost and EIO-LCA model	URS Corporation (2010), CMU GDI (2010)
Water consumption	Trucking emissions	Wang and Santini (2009), URS Corporation (2010),Stokes and Horvath (2006), PA DEP (2010)
Well completion:	If flaring, gas flow emission factor multiplied by flaring time	NY DEC (2009), PA DEP (2010)
Wastewater disposal:		
Deep well injection	Deep well injection costs and EIO-LCA model	US ACE (2006), CMU GDI (2010)
Production, processing, transmission and storage, and combustion	Assumed comparable to national average	Venkatesh *et al* (2011)

In section 3.1, we report point estimates of GHG emissions for a base case. In section **5**, we report range estimates and consider the sensitivity of point estimates to particular assumptions. Table **2** summarizes important parameter assumptions and possible ranges. Uniform or triangular distributions are assigned to these parameters based on whether we had two (uniform) or three (triangular) data points. When more data was available, parameters of probability distributions that best fit the data were estimated. A Monte Carlo analysis was performed using these distributions, to estimate the emissions from the various activities considered in our life cycle model.

Table 2: Parameter assumptions and ranges. (Note: sources for base case and range values are in table 1 and discussed in the supplementary material (available at stacks.iop.org/ERL/6/034014/mmedia).)

Parameter	Base case	Range
Area of access road (acres)	1.43	0.1–2.75
Wells per pad (number)	6	1–16
Area of well pad (acres)	5	2–6
Vertical drilling depth (ft)	8500	7000–10 000
Horizontal drilling length (ft)	4000	2000–6000
Fracturing water (MMgal/well)	4	2–6
Flowback fraction (%)	37.5	35–40
Recycling fraction (%)	45	30–60
Trucking distance between well site and water source (miles)	5	0–10
Trucking distance between well site and deep well injection facility (miles)	80	3–280
Well completion time with collection system in place (h)	18	12–24
Well completion time without collection system in place (days)	9.5	4–15
Fraction of flaring (%)	76	51–100
Initial 30 day gas flow rate (MMscf/day)	4.1	0.7–10
Average well production rate (MMscf/day)	0.3	0.3–10
Well lifetime (years)	25	5–25

Emissions from Marcellus Shale Gas Preproduction

Horizontal wells are drilled on a multi-well pad to achieve higher cost-effectiveness. It is reported that a Marcellus well pad might have as few as one well per pad and as many as 16, but more typically 6–8 (ICF International 2009, NY DEC 2009, Currie and Stelle 2010). As a base case scenario, we chose to analyze the typical pad with six wells, each producing 2.7 Bcf (3.0×10^9 MJ), representing an average of 0.3 MMcf per day of gas for 25 years. Other production estimates are higher. EQT (2011), for example, provides a production estimate of 7.3 Bcf (8.1 ×

10^9 MJ) and Range Resources at 4.4 Bcf (4.9×10^9 MJ) (Ventura 2009). Within the LCA framework the impacts are distributed across the total volume of gas produced during the lifetime of the well. Thus, the choice of using the low end ultimate recovery as the base case should be considered conservative. With Marcellus shale gas production currently in its infancy, the average production characteristics have significant uncertainty, so we perform an extensive sensitivity analysis over a range of flow rates and well lifetimes, as discussed below.

The EIO-LCA (CMU GDI 2010) model was used to estimate GHG emissions from the construction of the access road and the multi-well pad. These costs were estimated using the utility price cost estimation method (RSMeans 2005). The size of an average Marcellus well pad is reported as being between 2 and 6 acres and typically between 4 and 5 acres (16 000 and 20 000 m^2) during drilling and fracturing phase (NY DEC 2009, Columbia University 2009). The costs of constructing this pad are estimated to be \$3.0–\$3.3 million per well pad in 2002 dollars (see the supplementary information available at stacks.iop.org/ERL/6/034014/mmedia for detail). Using these costs as input, GHG emissions associated with well pad construction are estimated with the EIO-LCA (CMU GDI 2010) model.

Greenhouse gas emissions associated with drilling operations were calculated by two methods; (1) using the drilling energy intensity (table 1) and the life cycle diesel engine emissions factor of 635 g CO_2e per hp–hr output (Sheehan *et al* 2000), and (2) using drilling cost data and the EIO-LCA model (CMU GDI 2010). The EIA estimated the average drilling cost for natural gas wells in 2002 to be \$176 per foot (including the cost for drilling and equipping the wells and for surface producing facilities) (US EIA 2008). Emissions associated with the production of the drilling mud components were based on data from the SimaPro life cycle tool and the EIO-LCA economic model (PRé Consultants 2007, CMU GDI 2010).

Hydraulic fracturing associated GHG emissions result from the operation of the diesel compressor used to move and compress the fracturing fluid to high pressure, the emissions associated with the production of the hydraulic fracturing fluid, and from fugitive methane emissions as flowback water is captured. The last category of emissions is discussed separately below. Energy and emissions associated with the hydraulic fracturing process were modeled by using vendor specific diesel data along with the emission factor described above. The emissions of hydraulic fracturing fluid production are estimated with EIO-LCA model, based on the price of additives and fracturing fluid composition (see supplementary information available at stacks.iop.org/ERL/6/034014/mmedia for detail).

There may be significant GHG emissions as a result of flaring and venting activities that occur during well casing and gathering equipment installation. The natural gas associated with the hydraulic fracturing flowback water is flared and vented. Flaring is used for testing the well gas flow prior to the construction of the gas gathering system which transport the gas to the sales line. Well completion emissions depend on the flaring/venting time, gas flow rate during well completion, the ratio of flaring to venting, and flaring efficiency. Uncertainty/variability analysis was conducted to investigate the effect of flaring/venting time, gas flow rate during fracturing water flowback, and flaring per cent on the well completion emissions. For those well completions with the collection facilities in place, gas is flared for between 12 and 24 h, due to necessary flowback operations. In wells where the appropriate gas gathering system as a tie to the gas sales line is not available for the gas during fracturing water flowback, the flaring or venting can occur for between 4 and 15 days as shown in table 2 (NY DEC 2009). In our model, we assumed the gas release rate during well completion equals the initial 30 day gas production rate for the base case and considered a scenario with both venting and flaring (see supplementary information available at stacks.iop.org/ERL/6/034014/mmedia for details).

Emissions from Marcellus Shale Gas Production to Combustion

GHG emissions for production, processing, transmission, distribution and combustion of Marcellus shale natural gas are assumed to be similar to the US average domestic gas system that have been estimated previously (Jaramillo *et al* 2007). Jaramillo *et al* (2007) estimates were updated to include the uncertainty and variability in life cycle estimates and recalculated with recent and/or more detailed information by Venkatesh *et al* (2011). The GHG emissions from these life cycle stages consist of vented methane (gas release during operation), fugitive methane (unintentional leaks) and CO_2 emissions from the processing plants and from fuel consumption. Methane leakage rates throughout the natural gas system (excluding the preproduction processes previously discussed) are a major concern and our analysis has an implied fugitive emissions rate of 2%, consistent with the EPA natural gas industry study (US EPA 1996, 2010).

Venkatesh *et al* (2011) estimated the mean emission factors used in this study: 9.7 g CO_2e/MJ of natural gas in production; 4.3 g CO_2e MJ for processing; 1.4 g CO_2e/MJ for transmission and storage; 0.8 g CO_2e/MJ for distribution; and 50 g CO_2e/MJ for combustion.

Emissions Associated With the Life Cycle of Water Used For Drilling and Hydraulic Fracturing

Water resource management is a critical component of the production of Marcellus shale natural gas. Chesapeake Energy (2010) indicates that 100 000 gallons of water are used for drilling mud preparation. Two to six million gallons of water per well are required for the hydraulic fracturing process (Staaf and Masur 2009). About 85% of the drilling mud is reused (URS Corporation 2010). The flowback and recycling rates are used to estimate the total volume of water required. About 60–65% of this hydrofracturing fluid is recovered (URS Corporation 2010). For the flowback water, a recycle rate from 30 to 60% can be achieved (Agbaji *et al* 2009). The rest of the flowback water is temporarily stored in the impoundment and transported off site for disposal. Base case assumptions for these parameters are shown in table 2.

Emissions associated with drilling water use and hydraulic fracturing water use result from water taken from surface water resources or a local public water system; truck transport to the well pad, and then from the pad to disposal via deep well injection. It is assumed that no GHG emissions are related with producing water if it comes from surface water resources. For the water purchased from a local public water system, the emission factor for water treatment is used, which is estimated to be 3.4 g CO_2e/gallon of water generated according to Stokes and Horvath (2006). The energy intensity for transportation of liquids via truck is assumed to be 1028 Btu/ton mile for both forward and back-haul trips, as given in the GREET model (Wang and Santini 2009). In this study we assume that separate round trips are needed to transport the freshwater to the pad and to remove wastewater to the disposal site. This is to say that trucks bring in the freshwater from the source and return to the source empty; trucks also collect the wastewater from the well site and return to the well site empty. The life cycle emission factor (wells to wheels) for diesel as a transportation fuel is 93 g CO_2e/MJ (Wang and Santini 2009).

To estimate transport emissions associated with water taken from surface streams and water purchased from the local public water system, we used spatial analysis (ArcGIS) to estimate the distance from the surface water source to the well pad using well operational data and geographical information from Pennsylvania Department of Environmental Protection (2010). We depicted the overall distribution pattern of Marcellus wells under drilling and production in PA and NY in June 2010 by GIS. The distance from the well site to the surface water source is assumed to be 5 miles or 8 km in the base case of the model and the same transportation distance is also assumed for the water purchased from local public water system. We assumed an equal probability for sourcing water between surface water and the local public water system.

The trucking distance between well site and deep well injection facility was also estimated by GIS (PA DEP 2010). The average value of 80 miles or 130 km as determined by GIS was used in the base case.

RESULTS FOR THE BASE CASE

A total of 5500 t CO_2e is emitted during 'preproduction' per well. This is equivalent to 1.8 g CO_2e/MJ of natural gas produced over the lifetime of the well. Figure 3 depicts the GHG emissions by preproduction stage and by source. As can be seen, the completion stage has the largest GHG emissions, which result from flaring and/or venting. The error bars represent the limits of the 90% confidence interval of the emissions from each stage based on the uncertainty analysis.

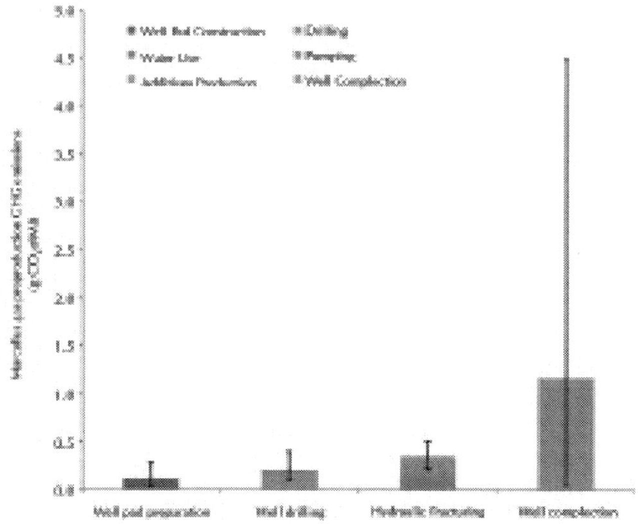

Figure 3: GHG emissions from different stages of Marcellus shale gas preproduction.

A recent EPA report addressing emissions from the natural gas industry reported that 177 t of CH_4 is released during the completion of an unconventional gas well (US EPA 2010). This estimate is consistent with the analysis here and falls within the range estimated by our study, 26–1000 t of CH_4 released per completion and a mean value of 400 t of CH_4 released per completion. In our model, this methane released during the well completion is either flared with a combustion efficiency of 98% or vented without recovery.

Adding the preproduction emissions estimate to the downstream emission estimated by Venkatesh *et al* (2011) results in an overall GHG emissions factor of 68 g CO_2e/MJ of gas produced (figure 4). The life cycle emissions are dominated by combustion that accounts for 74% of the total emissions.

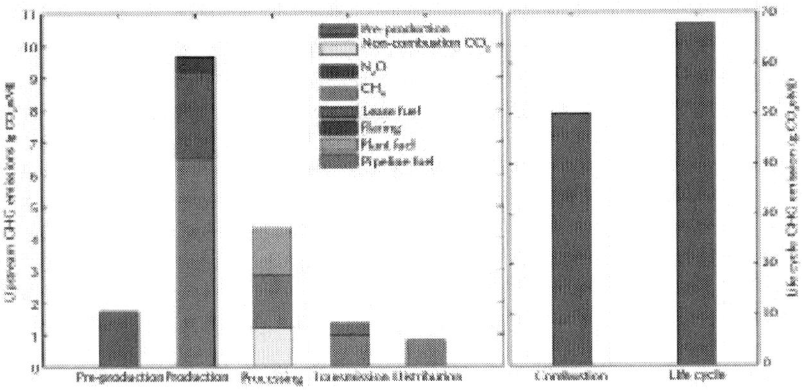

Figure 4: GHG emissions through the life cycle of Marcellus shale gas. (Pre-production through distribution emissions are on left scale; combustion and total life cycle emissions are on right scale. No carbon capture is included after combustion.).

SENSITIVITY AND UNCERTAINTY

Our results are subject to considerable uncertainty, particularly for the production rates and well lifetime. Table 3 summarizes the uncertainty analysis on the emission estimates for preproduction based on the distribution of parameters used.

Table 3: Uncertainty analysis on Marcellus gas preproduction

Life cycle stage	Mean (g CO_2e/MJ)	Standard deviation (g CO_2e/MJ)	COV	90% CI-L (%)	90% CI-U (%)
Well pad preparation	0.13	0.1	0.72	58	131
Drilling	0.21	0.1	0.50	51	95
Hydraulic fracturing	0.35	0.1	0.24	37	42
Completion	1.15	1.8	1.53	96	287
Total	1.84	1.8	0.96	67	179

Table 4 addresses model sensitivity to different estimates of ultimate gas recovery from wells, investigating the impact of different production rates and lifetimes. At high production rates and long well lifetimes the preproduction GHG emissions are normalized over higher volumes of natural gas than when using low flow rates and short well lifetimes. Comparing the case of 10 MMscf/day with a 25-year well lifetime to 0.3 MMscf/day with a 5-year well lifetime, table 4 shows that the emissions go from 0.1 to 9.2 g CO_2e/MJ. The overall life cycle emissions change from 65 to 74 g CO_2e/MJ. However, the preproduction emissions are less than 15% of the total life cycle emissions in all cases.

Table 4: Sensitivity of emissions from wells with different production rates and lifetimes. (Source: author calculations)

Average gas flow (MMscf/day)	**Lifetime (years)**	**Emissions from preproduction (g CO_2e/MJ)**	**Preproduction % contribution to life cycle emissions of Marcellus shale gas (%)**	**Total life cycle emissions (g CO_2e/MJ)**
10	25	0.1	0.1	65.3
10	10	0.1	0.2	65.3
10	5	0.3	0.4	65.5
3	25	0.2	0.3	65.4
3	10	0.5	0.7	65.7
3	5	0.9	1.4	66.1
1	25	0.6	0.8	65.8
1	10	1.4	2.1	66.6
1	5	2.8	4.1	68.0
0.3	25	1.8	2.7	67.0
0.3	10	5	6.6	69.8
0.3	5	9.2	12.4	74.4

COMPARISON WITH COAL FOR POWER GENERATION

Marcellus shale gas emissions can be compared to alternative energy sources and processes when using a common metric such as electricity generated. Currently coal power plants are used to generate base load. Natural gas power plants, especially inefficient ones, are used to provide regulation services to balance supply and demand at times when base load power plants are insufficient or there is high-frequency variability in load or from renewable resources. Natural gas combined cycle (NGCC) plants could be used to generate base load thus competing directly with coal to provide this service. For this reason our comparison includes the emissions associated with using Marcellus shale gas in a NGCC power plant (efficiency of 50%) and the emissions from using coal in pulverized coal (PC) plants (efficiency of 39%) and integrated gasification combined cycle (IGCC) plants (efficiency of 38%). The results of these comparisons can be seen in figure 5. For this comparison point values are used for the life cycle GHG emissions of coal-based electricity. The error bars found in figure 5 represent the low and high emissions values for Marcellus shale gas, based on the assumptions of well production rate and well lifetime. The high-emission scenario assumes a 5-year well with 0.3 MMscf/day production rate while the low-emission scenario, assumes a 25-year well with 10 MMscf/day production rate. Also shown in figure 5 are the life cycle emissions of electricity generated in power plants with carbon capture and sequestration (CCS) capabilities (efficiency of 43% for NGCC with CCS; efficiency of 30% for PC with CCS; efficiency of 33% for ICGG with CCS).

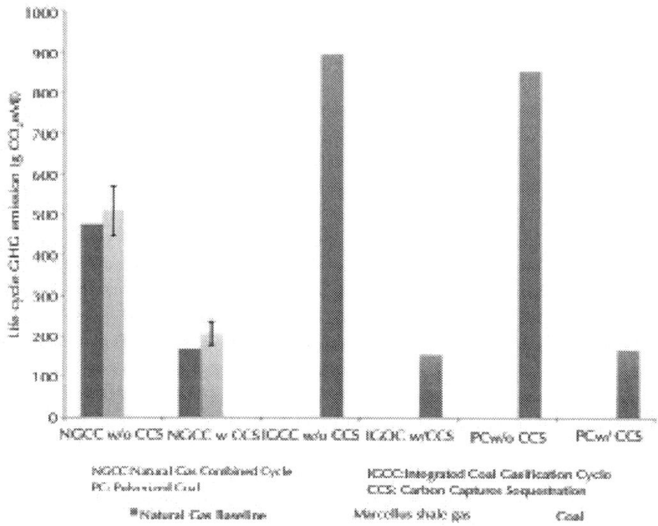

Figure 5: Comparison of life cycle GHG emissions from current domestic natural gas, Marcellus shale gas and coal for use in electricity production.

In general, natural gas provides lower greenhouse emission for all cases studied whether the gas is derived from Marcellus shale or the average 2008 domestic natural gas system. When advanced technologies are used with CSS then the emissions are similar and coal provides slightly less emissions. This implies that the upstream emissions for natural gas life cycle are higher than the upstream emissions from coal, once efficiencies of power generation are taken into account (Jaramillo *et al* 2007).

The comparison of natural gas and coal for electricity allows us to investigate the impact of three additional model uncertainty components including the choice of leakage rate, GWP values, and re-refracking of a Marcellus gas well. This study assumes a 2% production phase leakage rate based on the volume of gas produced (US EPA 2010, Venkatesh *et al* 2011). Assuming the average efficiency of 43% for natural gas fired electricity generation and 32% for coal fired plants the fugitive emissions rate would need to be 14% (resulting in a life cycle emission factor for Marcellus gas of 125 g CO_2e/MJ) before the overall life cycle emissions including those of electricity generation would be greater than coal. This is an exorbitantly high leakage rate and to put it into perspective, using 2009 dry natural gas production estimates and the average wellhead price, we calculate that the economic losses a would total around $11 billion. If we convert our data to the 20-year GWP the break-even point is reduced to 7% because of the higher impacts attributed to methane. Finally, we modeled a single hydraulic fracturing event occurring during well preproduction (figure 3). Above we calculated that the break-even emission factor that would make coal and natural electricity generation the same is 125 g CO_2e/MJ of natural gas. With the current emissions estimate for Marcellus gas of 68 g CO_2e/MJ, and a hydraulic fracturing event (and its associated flaring and venting emissions) contributing 1.5 g CO_2e/MJ to this estimate, more than 25 fracturing events would need to occur in a single well before the decision between coal and natural gas would change.

COMPARISON WITH LIQUEFIED NATURAL GAS AS A FUTURE SOURCE

In 2005 EIA suggested that domestic natural gas production and Canadian imports would decline as natural gas consumption increased. EIA predicted that liquefied natural gas (LNG) imports would grow to offset the deficits in North American production (US EIA 2011a, 2011b). As a result of the development of unconventional natural gas reserves, EIA has changed their projections. The Annual Energy Outlook 2011 reference case (US EIA 2011a, 2011b) predicts that increases in shale gas production, including Marcellus, will more than offset the decline in conventional natural gas and decreasing imports from Canada and will allow for increases in natural gas consumption. Since shale gas is projected to be the largest component of the unconventional sources of future natural gas production, it seem appropriate to compare its emissions to those of the gas that would be used if shale gas were not produced. Venkatesh *et al* (2011) estimated the life cycle GHG from LNG imported to the US to have a mean of 70 g CO_2e/MJ, These results are based on emissions due to production and liquefaction in the countries of origin, shipping the gas to the US by ocean tanker, regasification in the US and its transmission, distribution and subsequent combustion. On average, the emissions of Marcellus shale gas were about 3% lower than LNG. As with the overall Marcellus gas results, there is considerable uncertainty to the comparisons. However, we conclude that as these unconventional sources of natural gas supplant LNG imports, overall emissions will not rise.

CONCLUSIONS

The GHG emission estimates shown here for Marcellus gas are similar to current domestic gas. Other shale gas plays could generate different results considering regional environmental variability and reservoir heterogeneity. Green completion and capturing the gas for market that would otherwise be flared or vented, could reduce the emissions associated with completion and thus would significantly reduce the largest source of emissions specific to Marcellus gas preproduction. These preproduction emissions, however, are not substantial contributors to the life cycle estimates, which are dominated by the combustion emissions of the gas. For comparison purposes, Marcellus shale gas adds only 3% more emissions to the average conventional gas, which is likely within the uncertainty bounds of the study. Marcellus shale gas has lower GHG emissions relative to coal when used to generate electricity.

ACKNOWLEDGMENTS

We gratefully acknowledged the financial support from the Sierra Club. We also thank two anonymous reviewers and our colleagues Francis McMichael and Austin Mitchell for helpful comments. Any opinions, findings, and conclusions or recommendations expressed in this material are those of the authors and do not necessarily reflect the views of the Sierra Club.

REFERENCES

1. Agbaji A, Lee B, Kumar H, Belvalkar R, Eslambolchi S, Guiadem S and Park S 2009 Sustainable development and design of Marcellus shale play in Susquehanna PA: detailed report on the geology, geochemistry, technology and economics of the

Marcellus (available from http://ems.psu.edu/ ~ elsworth/courses/egee580/gas_final_2009.pdf, accessed October 2010)

2. Chesapeake2010WateruseinMarcellusdeepshalegasexploration (available fromhttp://chk.com/Media/MarcellusMediaKits/Marcellus_Water_Use_Fact_Sheet.pdf, accessed August 2010)
3. CMU GDI (Carnegie Mellon University Green Design Institute) 2010 Economic input–output life cycle assessment (EIO-LCA), US 1997 industry benchmark model (available from http://eiolca.net, accessed 27 August 2010)
4. Columbia University Urban Design Research Seminar 2009 *Hancock and the Marcellus Shale: Visioning the Impacts of Natural Gas Extraction Along the Upper Delaware* ed E Weidenhof *et al* (New York: The Open Space Institute Inc.) (available atwww.urbandesignlab.columbia.edu/sitefiles/file/HancockAndTheMarcellusShale.pdf, accessed August 2010)
5. CurrieKandStelleE2010Pennsylvania'snaturalgasboom:economic and environmental impacts *Policy Brief Commonwealth Found.* 22 (5) 1–11 (available from www.commonwealthfoundation.org/doclib/20100607_MarcellusShaleDrilling.pdf, accessed 30 July 2011)
6. EQT 2011 Investor relations (http://ir.eqt.com/index_d.cfm?Page = /index.cfm&pageName = Overview, accessed May 2011)
7. Harper J 2008 The Marcellus shale-an old 'new' gas reservoir in Pennsylvania *Pennsylvania Geol.* 38 2–13
8. Hendrickson C T, Lave L B and Matthews H S 2006 *Environmental Life Cycle Assessment of Goods and Services: An Input–Output Approach* (Washington, DC: Resources for the Future)
9. Hill D, Lombardi T and Martin J 2004 Fractured shale gas potential in New York *Northeastern Geol. Environ. Sci.* 26 57–78
10. ICF International 2009 *Technical Assistance for the Draft Supplemental Generic EIS: Oil, Gas and Solution Mining Regulatory Program*(Well permit issuance for horizontal drilling and high-volume hydraulic fracturing to develop the Marcellus Shale and other low permeability gas reservoirs—task 2) (Albany, NY: New York State Energy Research and Development Authority (NYSERDA)) (available at www.nyserda.org/publications/ICF%20task%202%20Report_Final.pdf, accessed 15 August 2010)

11. IPCC (Intergovernmental Panel on Climate Change) 2007 Climate Change 2007: Mitigation of Climate Change (Working Group III Report) (http://ipcc.ch/ipccreports/ar4-wg3.htm)
12. Jaramillo P, Griffin W M and Matthews H S 2007 Comparative life-cycle air emissions of coal, domestic natural gas, LNG, and SNG for electricity generation *Environ. Sci. Technol.* 41 6290–6
13. Kargbo D M, Wilhelm R G and Campbell D J 2010 Natural gas plays in the Marcellus shale: challenges and potential opportunities *Environ. Sci. Technol.* 44 5679–84
14. NY DEC (New York State Department of Environmental Conservation) 2009 *Draft Supplemental Generic Environmental Impact Statement (dSGEIS) for Oil, Gas and Solution Mining* (available from http://dec.ny.gov/energy/58440.html, accessed September 2010)
15. Osborn S G, Vengosh A, Warner N R and Jackson R B 2011 Methane contamination of drinking water accompanying gas-well drilling and hydraulic fracturing *Proc. Natl Acad. Sci.* 108 8172–6
16. PA DEP (Pennsylvania Department of Environmental Protection) 2010 *Oil and Gas Reporting*(http://marcellusreporting.state.pa.us/OGREReports/Modules/Welcome/Welcome.aspx, accessed November 2010)
17. PRé Consultants 2007 *SimaPro Version 7.1 Amersfoort* the Netherlands (available from http://pre.nl/, accessed August 2010)
18. RSMeans 2005 *Heavy Construction Cost Data* (Kingston, MA: RSMeans Company)
19. Shaker A S 2005 *Drilling Fluids Processing Handbook* (Burlington, MA: Gulf Professional Publishing) pp 319–24
20. Sheehan J, Camobreco V, Duffield J, Graboski M and Shapouri H 1998 An overview of biodiesel and petroleum diesel life cycles *NREL/TP-580-24772* (Golden, CO: National Renewable Energy Laboratory)
21. Soeder D J and Kappel W M 2009 Water resources and natural gas production from the Marcellus shale *US Geol. Survey Fact Sheet 2009–3032* (6pp)
22. Staaf E and Masur D 2009 Preserving forests, protecting waterways: policies to protect Pennsylvania's natural heritage from the threat

of natural gas drilling (available from http://PennEnvironment.org, accessed 30 July 2010)

23. Stokes J and Horvath A 2006 Life cycle energy assessment of alternative water supply systems *Int. J. Life Cycle Assess.* 11 335–43
24. Tilman D, Hill J and Lehman C 2006 Carbon-negative biofuels from low-input high-diversity grassland biomass *Science* 314 1598–600
25. URS Corporation 2010 *Water-Related Issues Associated with Gas Production in the Marcellus Shale: Additives Use Flowback Quality and Quantities Regulations* (Fort Washington, PA: New York State Energy Research and Development Authority (NYSERDA)) (available at www.nyserda.org/publications/urs-report-revised-po10666-2010.pdf, accessed 31 July 2011)
26. US ACE (US Army Corps of Engineers) 2006 *Cost Comparisons with the 1998 TSDFR (Treatment, Storage and Disposal Facilities)* (http://environmental.usace.army.mil/TSDF/rp0s12.htm, accessed May 2011)
27. US DOE (US Department of Energy) 2009 *Modern Shale Gas Development in the United States: a Primer* (Washington, DC: US DOE) (available from http://netl.doe.gov/technologies/oil-gas/publications/EPreports/Shale_Gas_Primer_2009.pdf, accessed 10 August 2010)
28. US EIA (US Energy Information Administration) 2008 *Costs of Crude Oil and Natural Gas Wells Drilled* (available fromhttp://eia.doe.gov/emeu/aer/txt/ptb0408.html, accessed 16 July 2010)
29. US EIA (US Energy Information Administration) 2011a *Summary Maps: Natural Gas in the Lower 48 States* (Washington, DC: US EIA) (available from http://www.eia.gov/oil_gas/rpd/shale_gas.jpg, accessed 26 July 2011)
30. US EIA (US Energy Information Administration) 2011b *Annual Energy Outlook 2011 with Projections to 2035* DOE/EIA-0383 (2011) (Washington, DC: US EIA)
31. US EPA (US Environmental Protection Agency) 1996 *Methane Emission from the Natural Gas Industry* EPA/600/SR-96/080 (Washington, DC: US EPA)

32. US EPA (US Environmental Protection Agency) 2010 *Greenhouse Gas Emissions Reporting from the Petroleum and Natural Gas Industry: Background Technical Support Document* (available from http://epa.gov/climatechange/emissions/downloads10/Subpart-W_TSD.pdf, accessed 12 December 2010)

33. Venkatesh A, Jaramillo P, Griffin W M and Matthews H S 2011 Uncertainty in life cycle greenhouse gas emissions from United States natural gas end-uses and its effects on policy (in preparation)

34. Ventura J 2009 Shale play comparison *Presented at the Developing Unconventional Gas Conference (DUG East) (Pittsburgh, PA, 19–20 October 2009)*

35. Wang M and Santini D 2009 *GREET (Greenhouse Gases, Regulated Emissions, and Energy Use in Transportation) Transportation Fuel Cycle Analysis Model Version 1.8* (Argonne, IL: Argonne National Laboratory) (available from http://greet.es.anl.gov/, accessed 7 November 2010)

Chapter 5

Environmental Public Health Dimensions of Shale and Tight Gas Development

Seth B.C. Shonkoff[1, 2], Jake Hays[3, 4], and Madelon L. Finkel[4]

[1]Physicians Scientists and Engineers for Healthy Energy, Oakland, California, USA

[2]Department of Environmental Science, Policy, and Management, University of California, Berkeley, Berkeley, California, USA

[3]Physicians Scientists and Engineers for Healthy Energy, New York, New York, USA

[4]Department of Public Health, Weill Cornell Medical College, New York, New York, USA

ABSTRACT

Background

The United States has experienced a boom in natural gas production due to recent technological innovations that have enabled this resource to be produced from shale formations.

Objectives

We reviewed the body of evidence related to exposure pathways in order to evaluate the potential environmental public health impacts of shale gas development. We highlight what is currently known and identify data gaps and research limitations by addressing matters of toxicity, exposure pathways, air quality, and water quality.

Discussion

There is evidence of potential environmental public health risks associated with shale gas development. Several studies suggest that shale gas development contributes to ambient air concentrations of pollutants known to be associated with increased risk of morbidity and mortality. Similarly, an increasing body of studies suggest that water contamination risks exist through a variety of environmental pathways, most notably during wastewater transport and disposal, and via poor zonal isolation of gases and fluids due to structural integrity impairment of cement in gas wells.

Conclusion

Despite a growing body of evidence, data gaps persist. Most important, there is a need for more epidemiological studies to assess associations between risk factors, such as air and water pollution, and health outcomes among populations living in close proximity to shale gas operations.

INTRODUCTION

Technological innovations in drilling and well-stimulation techniques have led to the production of natural gas from previously inaccessible geological formations, such as shale. Proponents of modern gas development argue that it has created a unique economic and political opportunity. Some in the public health community, however, have concerns about the potential for the extraction process to negatively impact the environment and human health (Finkel et al. 2013; Goldstein et al. 2012; Saberi 2013; Witter et al. 2013).

Producing natural gas from shale and tight gas formations in an economically feasible manner frequently requires a new constellation of existing technologies: high-volume, slickwater, hydraulic fracturing from clustered, multiwell pads using long directionally drilled laterals. This method can involve drilling a well vertically thousands of feet below the surface and then directionally (horizontally) for up to 2 miles. An average of 2–5 million gallons of fluid consisting of water, proppant (often crystalline silica), and chemicals (some of which are known carcinogens or otherwise toxic) are injected into the well at a pressure high enough to fracture the shale rock [U.S. Environmental Protection Agency (EPA) 2010a]. Chemicals often referred to as slickwater are added to the fracturing fluid to decrease its friction. The fracturing fluid creates and expands cracks in the shale. When the pressure is released, the cracks are held open by the sand, allowing the tightly held gases to flow into the cracks and up the production casing. The gas is then collected, processed, and sent through transmission pipelines to market. In 2012, shale gas constituted nearly 40% of U.S. gas production, up from 2% in 2000 (Hughes 2013).

Natural gas has a variety of attractive attributes. In the current market, it is a relatively inexpensive and abundant fuel. When combusted for electricity generation, it emits fewer health-damaging contaminants and approximately 50% less carbon dioxide emissions compared with burning coal (U.S. Energy Information Administration 2013). Yet, emerging scientific evidence suggests that there may be health risks associated with the development of shale gas.

In this review we discuss the body of scientific literature relevant to the environmental public health impacts of shale gas production. We highlight what is currently known and identify data gaps and research limitations.

METHODS

Scope of Review

For this review, we focused primarily on literature directly pertinent to the human health dimensions of shale and tight gas development. "Tight gas" refers to natural gas produced from reservoir rocks of low permeability, such as shale or sandstone. Shale gas and other forms of tight gas are referred to as "unconventional" because of their atypical reservoirs, which require new production techniques. However, we cite some studies that did not directly evaluate unconventional natural gas operations, but that are nonetheless relevant to various aspects of the overall process [e.g., particulate matter (PM) pollution, ozone]. In the case of ozone, for instance, we analyzed top-down studies that measured tropospheric concentrations rather than studies that supplied bottom-up measurements (e.g., leakage rates). Publications included in our review are predominantly sourced from the peer-reviewed scientific literature but include, where appropriate, government reports and other gray literature. Although the production chain of gas development is far reaching, we focused on the processes that begin with trucking the water, sand, chemicals, and other materials to the well pad, and end with the disposal of wastewater. Evidence suggests that these processes present the greatest risks to environmental public health and therefore have received the most attention in the scientific literature (Korfmacher et al. 2013; McKenzie et al. 2012; Rozell and Reaven 2012; Witter et al. 2013).

Terminology

Terminology is important when discussing modern forms of natural gas development. In part because of a lack of well-defined, uniform terminology, there has been confusion regarding which processes constitute this type of development. The terms, "hydraulic fracturing" or "fracking" are regularly used in the popular media as umbrella terms to describe the entire process of obtaining shale gas, as well as other forms of unconventional natural gas development, from land clearing and well spudding to transmission of natural gas to market. However, taken literally, "hydraulic fracturing" refers only to well-stimulation processes and excludes other potentially more health and environmentally impactful processes, including but not limited to well drilling, fracturing-fluid production, wastewater disposal, transportation of materials, and the processing, compression, and transmission of gas and liquids.

Many of the studies we cite in this review may also apply to other forms of oil and gas development that use well-stimulation techniques, including matrix acid stimulation, acid fracturing, and steam injection.

However, these other techniques are beyond the focus of this review. The term "unconventional oil and gas development" can also refer to bitumen/tar sands extraction and processing, and other types of fossil fuel development that employ novel engineering and production techniques to obtain fuels from unconventional resources (e.g., coalbed methane) that are beyond the scope of our review. Because most of the environmental public health–relevant scientific literature on modern oil and gas production has focused on the development of natural gas from shale formations, we use the term "shale gas development." However, here we discuss, where appropriate, scientific literature on other forms of unconventional or tight gas development that include the most prominent and relevant features of shale gas development, such as high-volume, horizontal hydraulic fracturing.

Identification of Relevant Studies

The literature directly relevant to the environmental public health dimensions of shale gas development is still limited. For this reason, we adopted a broad search strategy comprising the following:

- Systematic searches in three peer-reviewed science databases across multiple disciplines: PubMed (http://www.ncbi.nlm.nih.gov/pubmed/), Web of Science (http://www.webofknowledge.com), and ScienceDirect (http://www.sciencedirect.com)
- Searches in existing collections of scientific literature on this subject, such as the Marcellus Shale Initiative Publications Database at Bucknell University (http://www.bucknell.edu/script/environmentalcenter/marcellus), complemented by Google (http://www.google.com) and Google Scholar (http://scholar.google.com)
- Manual searches (hand-searches) of references included in all peer-reviewed studies that pertained directly to shale gas development.

For bibliographic databases, we used a combination of Medical Subject Headings (MeSH)-based and keyword strategies, which included the following terms, as well as relevant combinations:

shale gas, shale, hydraulic fracturing, fracking, drilling, natural gas production, Marcellus, Barnett, Denver-Julesberg Basin, air pollution, methane, water pollution, public health, water contamination, fugitive emissions, air quality, epidemiology, unconventional gas development, and environmental pathways.

This search identified a total of 211 peer-reviewed publications that pertain directly to shale gas development. (This database, the PSE STUDY CITATION DATABASE on Shale Gas & Tight Oil Development, is available online (http://psehealthyenergy.org/site/view/1180), and we will continue to update it with relevant literature.) Of these 211 publications, only 33 presented original data that met our inclusion criteria and that we considered relevant as primary literature.

Inclusion/exclusion criteria

From the studies identified through 1 February 2014, we excluded nonrelevant technical papers and studies related to economics, climate change, sociology, regulation, seismicity, water usage, social stress, and quality of life considerations. Although we excluded commentaries from the results of our review, a few are cited in order to provide documentation of particular considerations among the public health community. We included studies with direct pertinence to the environmental public health and environmental exposure pathways (i.e., air and water) associated with shale and tight gas development. In this regard, we supplemented the shale gas literature with studies that evaluated particular environmental pathways and health outcomes. For instance, we included studies directly related to the health impacts of tropospheric ozone, fine particulate air pollution, and endocrine-disrupting chemicals (EDCs). We excluded the vast majority of non–peer-reviewed scientific literature, but environmental impact statements and other government reports are cited where appropriate.

RESULTS

The environmental public health framework and possible exposure pathways

The environmental exposure pathway framework is often used to describe associations between pollutant sources and health effects via emissions, environmental concentrations of pollutants, pollutant exposure pathways (e.g., mouth, nose, ears, eyes, skin), and dose (i.e., micrograms of pollutant ingested per day) (Figure 1) (Agency for Toxic Substances and Disease Registry 2005).

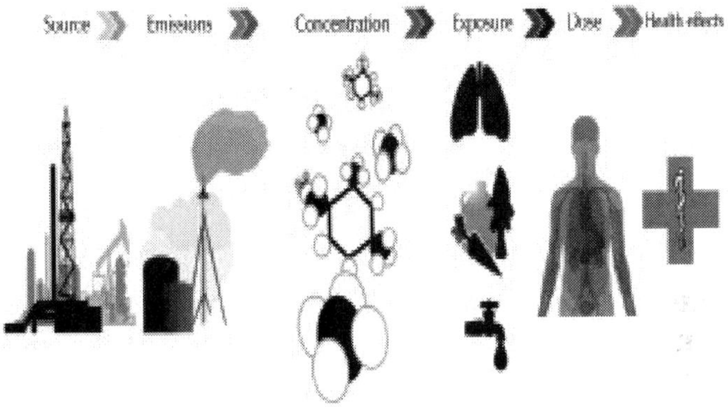

Figure 1: The environmental exposure pathway provides an analytical framework to describe, in broad terms, the connections between pollutant sources and human health outcomes. This framework begins with the emission source, in this case a well pad and associated infrastructure, which emit a variety of contaminants into the air, water, and soil. The concentrations of pollutants in the air, water, and soil that result from these emissions influence the magnitude of human exposures through organs such as the nose, mouth, and skin. Once the level of exposure is identified, it is then possible to estimate the dose, or how much of the pollutant is ingested in a given period of time. The dose, in turn, determines the health outcome.

Potential sources of health-relevant environmental pollution are present throughout many phases of shale gas development. These sources include shale gas production and processing activities (i.e., drilling, hydraulic fracturing, hydrocarbon processing and production, and wastewater disposal); the transmission and distribution of the gas to market (i.e., transmission lines and distribution pipes); and the transportation of water, sand, chemicals, and wastewater before, during, and after hydraulic fracturing.

HYDRAULIC FRACTURING FLUIDS: CHEMICAL TOXICOLOGY AND EXPOSURE PATHWAYS

Shale gas development uses fracturing fluids that contain organic and inorganic chemicals known to be health damaging (Aminto and Olson 2012; U.S. House of Representatives, Committee on Energy and Commerce 2011). Fracturing fluids can move through the environment and come into contact with humans in a number of ways, including surface leaks, spills, releases from holding tanks, poor well construction, leaks and accidents during transportation of fluids, flowback and produced water to and from the well pad, and run-off during blowouts, storms, and flooding events (Rozell and Reaven 2012). Further, the mixing of these compounds under conditions of high pressure—and often high heat—may synergistically create additional potentially toxic compounds (Kortenkamp et al. 2007;Teuschler and Hertzberg 1995; Wilkinson et al. 2000). Compounds found in these mixtures may pose risks to the environment and to public health through numerous environmental pathways, including water, air, and soil (Leenheer et al. 1982).

Chemicals are used in drilling and fracturing processes as corrosion inhibitors, biocides, surfactants, friction reducers, gels, and scale inhibitors, among others (Aminto and Olson 2012; New York State Department of Environmental Conservation 2011; Southwest Energy 2012). These chemicals include methanol, ethylene glycol, naphthalene, xylene, toluene, ethylbenzene, formaldehyde, and sulfuric acid, some of which are known to be toxic, carcinogenic, or associated with reproductive harm (Colborn et al. 2011; New York State Department of Environmental Conservation 2011). Many of these compounds are considered hazardous water pollutants and are regulated in other industries (Clean Water Act of 1972;Safe Drinking Water Act of 1974; U.S. House of Representatives 2011).

Many of the chemical compounds used in the fracturing process lack scientifically based maximum contaminant levels, making it more difficult to quantify their public health risks (Colborn et al. 2011). Moreover, uncertainty about the chemical makeup of fracturing fluids persists because of the limitations on required chemical disclosure, driven by the Energy Policy Act of 2005. For instance, in many states, companies are not mandated to disclose information about the quantities, concentrations, or identities of chemicals used in the process on the principle that trade secrets might be revealed (Centner 2013; Centner and O'Connell 2014; Maule et al. 2013).

Some companies make efforts to be more transparent in the disclosure of chemicals used in the process. FracFocus (http://www.fracfocus.org) was developed as an online, voluntary chemical disclosure registry, and some agencies (e.g., Bureau of Land Management) have suggested that it be used as a regulatory compliance tool (Konschnik et al. 2013). However, the registry has been criticized because of uncertainty surrounding the timing, substance, and omissions of the disclosed data on the website (Konschnik et al. 2013).

Because of the limited information that is available, researchers have sought to acquire more information on the chemical makeup of fracturing fluids through other means. For example, using material safety data sheets, Colborn et al. (2011) identified 353 of 632 chemicals contained in 944 products used for natural gas operations in Colorado, and they examined available information on each product. Their study represents one of the first attempts to conduct a chemical hazard assessment by identifying some of the compounds in fracturing fluids.

It should be noted that the scope of the study by Colborn et al. (2011) is limited in that they did not measure exposure, dose, or health outcomes across populations. The researchers identified Chemical Abstract Service (CAS) numbers for the chemicals and used these in systematic searches of databases such as TOXNET (http://toxnet.nlm.nih.gov). On the basis of search results, the researchers classified the compounds into 12 health-effects categories. At certain concentrations or doses, > 75% of the chemicals they identified are known to negatively impact the skin, eyes, and other sensory organs; the respiratory system; the gastrointestinal system; and the liver. Fifty-two percent of the chemicals have the potential to negatively affect the nervous system, and 37% are candidate EDCs (Colborn et al. 2011).

EDCs present unique hazards, particularly during fetal and early childhood growth and development (Diamanti-Kandarakis et al. 2009). They can affect the reproductive system and epigenetic mechanisms that may lead to pathology decades after exposure (Zoeller et al. 2012). EDCs have challenged traditional concepts in toxicology because effects at higher doses do not always predict effects at low doses (Vandenberg et al. 2012). In other words, the dose does not always make the poison.

Kassotis et al. (2014) measured estrogen and androgen receptor activity in surface and groundwater samples in Colorado using reporter gene assays in human cell lines. Water samples collected from the more intensive areas of natural gas development exhibited statistically significantly more estrogenic, antiestrogenic, or antiandrogenic activity than references sites with either no operations or fewer operations (Kassotis et al. 2014). The concentrations of chemicals detected were in high enough concentrations to interfere with the response of human cells to male sex hormones and estrogen. This study by Kassotis et al. (2014) indicated that EDCs are a potential health concern in natural gas operations, and suggested that chemicals used in the process should be screened for EDC activity.

AIR QUALITY

Air pollutant emission sources from shale gas development can be grouped into two main categories: *a*) emissions from drilling, processing, well completions, servicing, and other gas production activities; and *b*) emissions from transportation of water, sand, chemicals, and equipment to and from the well pad.

Air pollution: drilling, well stimulation, gas production, processing, and servicing

The literature suggests that shale gas development processes emit hazardous air pollutants including—but not limited to—BTEX compounds (benzene, toluene, ethylbenzene, and xylene), formaldehyde, hydrogen sulfide, acrylonitrile, methylene chloride, sulfuric oxide, nitrogen oxides (NO_x), volatile organic compounds (VOCs), trimethylbenzenes, aliphatic hydrocarbons, diesel PM, and radon gas (McKenzie et al. 2012; Pétron et al. 2012; Roy et al. 2013). These emissions can result in elevated air pollution concentrations that exceed U.S. EPA guidelines for both carcinogenic and noncarcinogenic health risks (McKenzie et al. 2012;Meteorological Solutions Inc. 2011).

McKenzie et al. (2012) used U.S. EPA guidance to estimate chronic and subchronic non-cancer hazard indices (HIs) and cancer risks from exposure to hydrocarbons for residents living > 0.5 mile and ≤ 0.5 mile from wells in Colorado. The authors found that residents living ≤ 0.5 mile from wells were at a greater risk for health effects from exposure to natural gas development than those living > 0.5 mile from wells. Notably, they found a subchronic non-cancer HI of 5 for those living ≤ 0.5 mile compared with an HI of 0.2 for those living > 0.5 mile from wells, which was driven primarily from exposure to trimethylbenzenes, xylenes, and aliphatic hydrocarbons (McKenzie et al. 2012). Unfortunately, baseline air quality data prior to the study were not available. However, the statistically significant spatial associations between air quality and shale gas development indicate that air quality may be negatively impacted and health risks may increase during various stages of shale gas development.

Bunch et al. (2013), however, found that shale gas production activities did not result in community-wide exposures to concentrations of VOCs at levels that would pose a health concern. They compared VOC concentration data from seven air monitors at six locations in the Barnett Shale region in Texas with federal and state health-based air comparison values (HBACVs) in order to determine possible acute and chronic health effects; none of the concentrations exceeded acute HBACVs (Bunch et al. 2013). Air quality data included in their study were generated from monitors focused on regional atmospheric concentrations of pollutants (Bunch et al. 2013). Conversely, McKenzie et al. (2012) included samples at the community level in close proximity to gas development. Finer geographically scaled samples often capture local atmospheric concentrations that are more relevant to human exposure. This may be a primary reason why health hazard estimates differed between the two studies.

Roy et al. (2013) estimated emissions of NO_x, VOCs, and PM for an air emissions inventory for the development of natural gas in the Marcellus Shale region for 2009 and 2020. They predicted that, in 2020, shale gas development activities would contribute 6–20% (mean, 12%) of the NO_x emissions and 6–31% (mean, 12%) of anthropgenic VOC emissions in Pennsylvania. However, these estimates were based on assumptions of improvements in gas production, completion, and processing infrastructure. If source-level emissions remain the same as in 2009, Marcellus VOC emissions were predicted to constitute approximately 34% (19–62%) of the regional anthropogenic VOC emissions in 2020 (Roy et al. 2013). Increases in emissions of VOCs and NOx, which are precursors of tropospheric ozone formation, could complicate ozone management in the region and potetentially offset ozone precursor emission reductions in other sectors at a time when several regions in Pennsylvania struggle to be within ozone attainment (Roy et al. 2013).

In another study focused on hydrocarbon emissions, Colborn et al. (2014) assessed air quality in western Colorado using weekly air samples collected before, during, and after drilling and hydraulic fracturing on a new natural gas well pad. They found numerous chemicals in the air samples that are associated with natural gas development operations, most notably methane, ethane, propane, and other alkanes. Many non-methane hydrocarbons (NMHCs), which were observed during the initial drilling phase, are associated with multiple health effects. Notably, 30 of the NMHCs they observed in the field were EDCs. In addition to the direct air pollution associated with natural gas drilling and processing (e.g., NMHCs, VOCs) outlined above, there are also indirect pollution concerns, such as the secondary atmospheric formation of tropospheric (ground-level) ozone (Colborn et al. 2014).

Studies have indicated that shale gas development is associated with the production of secondary pollutants such as tropospheric (ground-level) ozone, which is formed through the interaction of methane, VOCs, and NO_x in the presence of sunlight (Jerrett et al. 2009; U.S. EPA 2013). Tropospheric ozone is a strong respiratory irritant associated with increased respiratory and cardiovascular morbidity and mortality (Jerrett et al. 2009; United Nations Environment Programme 2011). Although toxicological data suggest that pure methane is not by itself health damaging (excluding its role as an asphyxiant and an explosive), it is a precursor to global tropospheric ozone (Smith et al. 2009).

Pétron et al. (2012) analyzed data collected at the National Oceanic and Atmospheric Administration (NOAA) Boulder Atmospheric Observatory (http://www.esrl.noaa.gov/psd/technology/bao) and filtered by wind sector, which indicated a high alkane and benzene signature from the direction of the Denver-Julesburg Basin, an area of considerable oil and gas development. The researchers found that an estimated 4% (range, 2.3–7.7%) of all natural gas (composed mostly of methane) produced was being accidentally leaked or purposefully vented to the atmosphere (Pétron et al. 2012).Karion et al. (2013) observed significant methane leaks in the Uintah Basin shale gas field, with an estimated 6.2–11.7% of total gas production leaking into the atmosphere.

In a national methane emissions study that combined ground and aerial sampling of methane with computer modeling, Miller et al. (2013) found that atmospheric levels of methane due to oil and gas extraction could be 4.9 ± 2.6 times greater than current estimates from the Emissions Database for Global Atmospheric Research (EDGAR) (http://edgar.jrc.ec.europa.eu/index.php) and the U.S. EPA. Although it is difficult to distinguish the sources of methane between oil and gas production and gas production, transmission, and storage, Peischl et al. (2013) estimated that 17% of gross methane production from oil and gas activities in the Los Angeles Basin are leaked or vented to the atmosphere.

Some studies have modeled ozone impacts associated with shale gas operations. Kemball-Cook et al. (2010) modeled ozone precursor emissions (VOCs and NO_x) in the Haynesville Shale play that lies beneath the northeast Texas/northwest Louisiana border. Photochemical modeling for 2012 showed increases in 8-hr ozone design values of up to 5 ppb, which, along with the amount of projected emissions, give cause for concern about future atmospheric concentrations of ozone in Texas and Louisiana (Kemball-Cook et al. 2010).Olaguer (2012) used the Houston Advanced Research Center (HARC) neighborhood air quality model to simulate ozone formation near a hypothetical natural gas–processing facility, using estimates based on both regular and nonroutine (e.g. flaring) emissions. This model predicted that under average conditions, using regular emissions associated with compressor engines may significantly increase ambient ozone in the Barnett Shale formation (> 3ppb at 2 km downwind from the facility) (Olaguer 2012).

Substantial air quality impacts from oil and natural gas operations in Wyoming, Colorado, Utah, and Texas have also been directly measured (Carter and Seinfeld 2012; Edwards et al. 2013; U.S. Department of Energy 2011). Schnell et al. (2009) studied air quality in the rural Upper Green River Basin (UGRB) of Wyoming near the Jonah-Pinedale Anticline natural gas field in February 2008. They observed high photochemical ozone concentrations in the UGRB in the winter, reporting readings of up to 140 ppb, just less than double the U.S. EPA ozone concentration limit of 75 ppb (U.S. EPA 2012a). Before 2005, typical wintertime ozone concentrations in this area were 30–40 ppb (Pinto 2009). This increase in ozone concentration during this time period could be associated with the increase in NO_x and VOC emissions from oil and gas development activities in the area (Schnell et al. 2009). In a study conducted for the Wyoming Department of Environmental Quality, Meteorological Solutions Inc. (2011) found that the 8-hr ozone concentrations in the UGRB in 2011 exceeded the U.S. EPA ozone 8-hr standard for 13 days (Meteorological Solutions Inc. 2011) and exceeded the U.S. EPA scientists-recommended limit of 65 ppb for 25 days (Weinhold 2008).

In Utah there were 68 days in the winter of 2010 when ozone levels exceeded the U.S. EPA ozone standard of 75 ppb, and in 2011 there were readings more than double the U.S. EPA standard (Utah Department of Environmental Quality 2013). Results of experiments conducted by the U.S. EPA and NOAA indicated that ozone precursor emissions (VOCs and NO_x, primarily) from oil and gas development in the Uintah Basin in Utah were a primary factor in the increased ozone level (Utah Department of Environmental Quality 2013).

Crystalline silica sand, used as a proppant (to "prop" open cracks in the target formation to allow gas to flow up the well), is delivered by trucks to the drilling site. Transporting this sand in trucks and trains and mixing it into fracturing fluids with sand movers, conveyer belts, and blender hoppers at the well site release silica dust into the air, where well-pad workers can be exposed (Esswein et al. 2013). Workers experience the most direct exposure; however, silica dust may also be an air contaminant of concern to nearby residents. The etiological association between respiratory exposure to silica dust and the development of silicosis is well known [Centers for Disease Control and Prevention (CDC) 1992, 2002]. Silicosis is a progressive lung disease in which tissue in the lungs reacts to silica particles, and can result in inflammation and scarring, which decreases the ability of the lungs to take in oxygen (CDC 1992, 2002). Respiratory exposure to silica is also associated with other diseases such as chronic obstructive pulmonary disease, tuberculosis, kidney disease, autoimmune conditions, and lung cancer (CDC 2002).

In cooperation with industry partners, Esswein et al. (2013) collected full-shift air samples at 11 sites in five states to determine levels of worker exposure. Of 111 air samples, 51.4% showed silica exposures greater than the calculated Occupational Safety and Health Administration permissible exposure level and 68.5% showed exposures greater than the National Institute for Occupational Safety and Health recommended exposure limit of 0.05 mg/m^3 (Esswein et al. 2013). Further, these researchers noted that the type of respirators worn by workers were not sufficiently protective in some cases, given the magnitude of silica concentrations (Esswein et al. 2013).

Air pollution: transportation

Each well requires on average between 2 and 5 million gallons of water per hydraulic fracturing event (U.S. EPA 2010a). Water is generally not pumped directly to wells but is instead transported by diesel trucks, each of which has an approximate capacity of 3,000 gallons (U.S. EPA 2011b). It has been estimated that approximately 2,300 trips by heavy-duty trucks are required for each horizontal well during early stages of shale gas development (New York State Department of Environmental Conservation 2011). With thousands of such wells concentrated in high-development regions, levels of truck traffic and diesel-associated air pollution will increase in these areas.

The pollutant of primary health concern emitted from the transportation component of shale gas development is fine diesel PM. Diesel PM is a well-understood health-damaging pollutant that contributes to cardiovascular illnesses, respiratory diseases (e.g., lung cancer) (Garshick et al. 2008), atherosclerosis, and premature death (Pope 2002; Pope et al. 2004). For example, a study from the California Air Resources Board (Tran et al. 2008) indicated that there is an expected 10% (uncertainty interval: 3%, 20%) increase in the number of premature deaths per 10-μg/m^3 increase in $PM_{2.5}$ ($PM \leq 2.5$ μm in aerodynamic diameter) exposure. Particulates can also contain concentrated associated products of incomplete combustion, and when particle diameter is < 2.5 μm, they can act as a delivery system to the alveoli of the human lung (Smith et al. 2009). In addition to diesel PM, as previously mentioned, NO_x and VOCs—other pollutants prevalent in diesel emissions—react in the presence of sunlight and high temperatures to produce tropospheric (ground-level) ozone.

WATER QUALITY

Rozell and Reaven (2012) conducted a risk assessment that identified five main pathways of water contamination in the shale gas production process: *a*) transportation spills of fracturing fluid or produced water; *b*) well casing leaks; *c*) leaks through fractured rock; *d*) drilling site discharge; and *e*) wastewater disposal. They found that wastewater disposal carries a potential risk of water contamination several orders of magnitude larger than that of the other pathways (Rozell and Reaven 2012).

Other studies have suggested that structural impairment of cement used to prevent transzonal gas migration in the wellbore is the most common mechanism through which groundwater can become contaminated (Vidic et al. 2013). Indeed, state environmental regulators at the Pennsylvania Department of Environmental Protection found that oil and gas development was responsible for polluting water supplies for at least 161 residences in Pennsylvania between 2008 and 2012, primarily due to cement structural integrity in wells and wellbores (Legere 2013). For the purpose of this review, we focused primarily on well casing leaks, drilling site discharge, and wastewater disposal because these are generally regarded as the most viable means of water contamination (Rozell and Reaven 2012; Vidic et al. 2013).

Flowback and produced water

Estimates of the proportion of fracturing fluid that returns to the surface as flowback and produced waters range from 9% to 80%, with most estimates around 35% (Horn 2009; New York State Department of Environmental Conservation 2011;U.S. EPA 2010a). These wastewaters contain the chemicals used in the fracturing fluid as well as compounds found deep in geological strata, such as salts, chlorides, heavy metals (e.g., cadmium, lead, arsenic), organic chemicals (e.g., BTEX compounds), bromide, and—depending on the geology—naturally occurring radioactive materials (e.g., radium-226). Many of these naturally occurring compounds are associated with human health effects when exposure is sufficiently elevated (Balaba and Smart 2012; Colborn et al. 2011; Haluszczak et al. 2013). A proportion of flowback and produced waters are treated and released as effluent or for other beneficial uses, such as irrigation for agriculture. However, many of the chemicals persist in high quantities because treatment facilities are unable to screen for and eliminate the complex array of compounds and products of synergistic interactions among them (Ferrar et al. 2013; Hladik et al. 2014; Lutz et al. 2013).

Flowback and produced water are sometimes treated at facilities and then discharged into surface waters (Ferrar et al. 2013). Warner et al. (2013a) examined water quality and isotopic compositions of discharged effluents, surface waters, and stream sediments associated with a Marcellus wastewater treatment facility site. Their findings suggest that insufficiently treated flowback and produced water that contained elevated concentrations of contaminants associated with shale gas development entered local water supplies, even after treatment. They also found elevated levels of chloride and bromide downstream, along with radium-226 levels in stream sediments at the point of discharge, that were approximately 200 times greater than upstream and background sediments and well above regulatory standards (Warner et al. 2013a). These types of water emissions may increase the health risks of residents who rely on these surface and hydrologically contiguous groundwater sources for drinking water (Wilson and VanBriesen 2012) and sources of food (i.e., fish protein) (Papoulias and Velasco 2013).

In a meta-analysis of chemical and physical characterizations of produced waters from shale gas, Alley et al. (2011) found that most of the produced waters generated by shale gas development were classified as saline (> 30,000 mg/L) or hypersaline (> 40,000 mg/L). These authors pointed out that treatment of this produced water for beneficial use often involves reverse osmosis, a practice that may generate a waste stream too large to justify the activity.Alley et al. (2011) also found that prior to treatment, produced waters can exceed toxicity thresholds of contaminants of concern, including—but not limited to—phosphates, cadmium, aluminum, barium, chloride, strontium, radium-226, bromine, lithium, and magnesium. Toxicity thresholds used in their meta-analysis were LC_{50} values (concentration lethal to 50%) for *Ceriodaphnia dubia* Richard, *Daphnia magna* Straus, and *Pimephales promelas*Rafinesque, and water-use criteria from the Food and Agricultural Organization of the United Nations guidelines for agricultural uses and the U.S. EPA Water Quality Criteria for surface discharge (Alley et al. 2011).

The results of Alley et al. (2011) agree with other reports that samples of fracturing fluids, drilling muds, and flowback and produced waters in wastewater–surface containment ponds contain chemicals that, at elevated doses or certain concentrations, have been associated with health effects ranging from skin and eye irritation to neurological and nervous system damage, cancer, and endocrine disruption (Colborn et al. 2011). Moreover, between July 2009 and June 2010, 192.5 million gallons of produced water was reported in Pennsylvania alone, with no certainty as to the location and type of disposal to be employed (Pennsylvania Department of Environmental Protection 2010).

The handling and disposal of flowback and produced water also hold implications for air quality because of volatile compounds, such as BTEX compounds, that are often mixed with the fluids. This may be particularly relevant when wastewater is stored in surface containment ponds and misted into the air to promote evaporation (Colborn et al. 2011).

Gas and fluid migration

Subsurface gas and fluid migration is most commonly associated with impaired structural integrity of well cement and, to a lesser extent, well casings. Failures in well barriers may allow intrusion of gases and fluids from producing formations below the casing shoe or from shallower gas- and fluid-bearing formations intersected by the wellbore to lower-pressure annuli. This may result in annular gas flow or sustained casing pressure and thus become a pathway for gas migration to the surface, a known mechanism of emissions of gases to the air and migration of gases and fluids to groundwater (Brufatto et al. 2003; Watson and Bachu 2009). Methane and other hydrocarbons can also migrate along improperly plugged wells, through an inadequately sealed annulus, or between geological zones as a result of cement failures in the wellbore (Vidic et al. 2013).

Leaking oil and gas wells have been recognized as a potential mechanism of subsurface migration of methane and heavier *n*-alkanes and other non-methane VOCs into groundwater and the atmosphere, contributing risks to drinking water and air quality (Bourgoyne et al. 2000; Brufatto et al. 2003; Chilingar and Endres 2005; Watson and Bachu 2009). Cement failures in onshore and offshore wells, reported to occur in 2–50% of all wells, provide pathways for gas migration to occur in the wellbore (Bourgoyne et al. 2000; Brufatto et al. 2003; Watson and Bachu 2009).

Because methane has a low solubility (26 mg/L at 1 atm, 20°C) (Vidic et al. 2013) and is relatively unreactive compared with longer-chain and unsaturated hydrocarbons (Jackson et al. 2011), it is typically regarded as nontoxic and is not regulated in the United States as a solute in water wells. However, there are no peer-reviewed studies on the health effects of chronic exposure to lower concentrations of methane in drinking water or indoor or outdoor air (Jackson et al. 2011). Further, if there is a pathway for methane migration, there could be a pathway for associated health-damaging gases coproduced with methane.

Some attention has been paid to the flammability of methane, the risk of explosions, and the risk of asphyxiation (primarily in high indoor concentrations). For example, in 2007 in Geauga County near Cleveland, Ohio, methane contaminated a water well and a home exploded; the Ohio Department of Natural Resources blamed a faulty concrete casing in a nearby gas well (Ohio Department of Natural Resources 2008). Similarly, in Pavillion, Wyoming, high concentrations of methane found in drinking water wells were attributed to gas production activities (DiGiulio et al. 2011). In addition, the U.S. EPA concluded that methane from geological layers not targeted for gas production migrated up the wellbore to an aquifer as a result of well cement failures in Parker County, Texas (U.S. EPA 2010b).

In certain regions, methane can naturally occur in aquifers, and there are conflicting scientific opinions about whether its presence is caused or exacerbated by shale gas development (Davies 2011; Saba and Orzechowski 2011; Schon 2011). However, there are convincing findings that shed light on the likelihood that shale gas development is associated with high methane levels in drinking water wells. Osborn et al. (2011) found that communities in Pennsylvania that had active shale gas development (one or more gas wells within 1 km) had statistically significantly higher concentrations of methane in their water wells compared with nonextraction sites (no shale gas wells within 1 km). The chemical signature of the methane found in drinking water wells in the active area indicated that the methane came from a high-pressure, deep-earth source (thermogenic methane). Alternatively, the methane from nonactive sites had signatures of shallow earth origins (biogenic methane). This suggests that shale gas production processes were the source of the methane contamination (Osborn et al. 2011).

Building on previous work by Osborn et al. (2011), Jackson et al. (2013) analyzed 141 drinking water wells across northeastern Pennsylvania. The researchers found methane in 82% of the samples (115 of 141 wells), with average concentrations six times higher for homes that were < 1 km from natural gas wells (59 of 141). These data, based on isotopic signatures and gas ratios, suggest that a subset of homeowners living < 1 km from shale gas wells had drinking water that was contaminated with stray gases associated with gas development activities (Jackson et al. 2013).

There is evidence that, in some locations, pathways exist between deep underlying formations and shallow drinking water aquifers (Vengosh et al. 2013). A modeling study byMyers (2012) suggested that pathways would allow for the transport of contaminants from the fractured shale to aquifers. Warner et al. (2012) found evidence of possible migration of Marcellus brine through naturally occurring pathways, based on strong geochemical fingerprints in salinized groundwater samples.

Both of these studies (Myers 2012; Warner et al. 2012) suggest that migration through fractured rock can serve as a subsurface contamination pathway to underground sources of drinking water. They also highlight the significance of the specific geographic configuration because some shallow drinking water resources are at more risk for contamination than others. In a study of the Fayetteville Shale in Arkansas, Warner et al. (2013b) suggested that methane contamination of shallow groundwater may not be a problem in certain shale formations. This difference may be attributed to geological variations across geographic space, including the presence of intermediate gas-bearing formations that are found overlying parts of some shale plays (e.g., Marcellus) but not others (e.g., Fayetteville).

In addition, Fontenot et al. (2013) evaluated water quality in private drinking water wells near natural gas operations in the Barnett Shale formation in Texas and found higher levels of arsenic, selenium, strontium, and total dissolved solids in wells located within 3 km of active gas wells. The authors used historical data from the region as a baseline to determine the contamination rates before the expansion of natural gas operations. Although heavy metals were present at low levels in aquifers in the region, concentrations were significantly higher in areas of active development (Fontenot et al. 2013). The authors were able to link contamination to natural gas activities; however, the specific factor responsible for contamination (e.g., well casing failures, mobilization of natural constituents, hydrogeochemical changes from lowering the water table) was not determined (Fontenot et al. 2013).

Researchers have been challenged in their ability to link associations between water contamination and unconventional natural gas development to any particular part of the process. After complaints about the taste and odor of well water from residents of Pavillion, Wyoming, the U.S. EPA initiated a groundwater investigation (DiGiulio et al. 2011). The observed water wells were located in an area known as the Pavillion gas field, which contained 169 gas production wells and 33 containment ponds used for storage/disposal of drilling wastes and produced and flowback waters from unconventional natural gas development of a sandstone formation.

From 2009 to 2011 the U.S. EPA conducted four sampling events meant to determine the presence (not extent) of groundwater contamination in the formation. In that study, DiGiulio et al. (2011) detected elevated concentrations of benzene, toluene, ethylbenzene, and xylenes (BTEX) in sampling wells at concentrations of 246, 617, 67, and 750 μg/L, respectively. Trimethylbenzenes and diesel range organics were detected at concentrations up to 105 and 4,050 μg/L, respectively, and total purgeable hydrocarbons were detected in the groundwater samples near the containment ponds (DiGiulio et al. 2011). Although these initial data indicated groundwater impacts that seem likely to be associated with unconventional gas production practices (U.S. EPA 2011a), the results of the study byDiGiulio et al. (2011) have been contested, and it is still unclear which part of the gas development process (if any) is responsible for the contamination. Further, there are geological differences between sandstone and shale, and fracturing is often conducted closer to the surface in sandstone formations. However, the findings suggest an association between water contamination and production activities that have also been identified in shale gas development (DiGiulio et al. 2011).

Site discharge and improper waste disposal

Fracturing fluids and produced waters can also contaminate underground sources of drinking water during waste management and disposal. Flowback and produced waters are often contained in evaporation ponds, pits, and tanks, in some cases in very close proximity to residences (Bamberger and Oswald 2012; Rozell and Reaven 2012). These containment ponds are often, but not always, lined to protect against leakage; however, case studies have documented reported ruptures to these liners that may have led to water and soil contamination and contributed to fish and livestock deaths (Bamberger and Oswald 2012). An analysis of waste obtained from reserve pits indicated the potential for exposure to technologically enhanced naturally occurring radioactive material and potential health effects from individual radionuclides (Rich and Crosby 2013).

Groundwater contamination can also result from surface spills at active well sites. Gross et al. (2013) analyzed data from the Colorado Oil and Gas Conservation Commission (http://cogcc.state.co.us) and noted 77 reported surface spills (associated with < 0.5% of active wells) impacting groundwater in Weld County, Colorado. The groundwater samples were analyzed for BTEX components. Most notably, benzene measurements exceeded the U.S. EPA National Drinking Water maximum contaminant level of 5 ppb in 90% of the samples (Gross et al. 2013). Because baseline-sampling measurements were not available, the background BTEX concentrations remain unclear. However, natural groundwater concentrations are typically low near deposits of crude oil, coal, and natural gas (Gross et al. 2013).

DISCUSSION

Future research needs

There is a growing body of scientific literature on the environmental public health dimensions of shale gas development; however, a number of important data gaps persist. Measurements of emissions and atmospheric concentrations should be conducted among diverse geographies, both indoors and outdoors, to help to estimate the types and magnitude of population exposures to pollutants associated with shale gas development. In addition, studies that take into account personal exposures and time–activity patterns of individuals would be helpful in assessing epidemiologically meaningful exposures. These studies could include the use of personal monitors and sampling of household drinking water in conjunction with health records to look at disease outcomes.

Perhaps the most important information gap is the lack of epidemiological studies. There is a need to assess the strength of the association between risk factors, such as air pollution and water contamination, and health outcomes among populations living in close proximity to shale gas develoment activities compared with those populations living in areas without these activities. Although lacking in definitive proof of cause and effect, self-reporting health surveys and environmental testing have suggested possible adverse health outcomes from shale gas development in Pennsylvania (Steinzor et al. 2013). Of particular interest are the epidemiological studies on vulnerable populations, including pregnant women, young children, the elderly, and those with compromised immune systems, who live, work, and play in close proximity to shale gas development. Because workers are likely to be the first and the most exposed demographic from shale gas development, further occupational health studies are also needed.

There have been some efforts in epidemiology and risk assessment, including a recent retrospective cohort study by that examined associations between maternal residential proximity to natural gas development and a number of birth outcomes. The authors found no positive association between density and proximity of wells within a 10-mile radius of maternal residence and prevalence of oral clefts, preterm birth, or term low birth weight. However, the researchers did observe a positive association between density and proximity of pregnant mothers to shale gas development and the prevalence of congenital heart defects and possibly neural tube defects in their newborns (McKenzie et al. 2014).

There have been some other epidemiological efforts as well, including a study funded by America's Natural Gas Alliance that evaluated associations between childhood cancer incidence in Pennsylvania and hydraulic fracturing sites (Fryzek et al. 2013). The authors included

29,000 hydraulically fractured wells drilled between 1990 and 2009 in their analysis and obtained data on childhood cancers from the Pennsylvania cancer registry for this time period. However, shale gas development did not begin in Pennsylvania until 2006, when four wells of this type were drilled. In fact, only 726, or 2.5% of the 29,000 wells in their database, were relevant to directionally drilled shale gas wells. Unfortunately, this exposure misclassification and the disregard for the extended latency periods of many childhood cancers render this study inconclusive as to the effect of shale gas development on childhood cancer rates. The study by Fryzek et al. (2013) demonstrates the need for more epidemiological assessments that pay attention to the latency periods of environmentally mediated diseases.

Epidemiological investigations are challenged by the difficult task of identifying specific risk factors and the uncertainty in exposure classification because compounds used in shale gas development are often not disclosed. In these cases of uncertainty, a comprehensive water monitoring and—under certain circumstances, a biomonitoring program—that uses both targeted and nontargeted strategies would be useful. Useful data could be generated by targeted testing for specific compounds known to be associated with shale gas development in drinking water supplies and in the blood and urine of a representative sample of individuals living in close proximity to shale gas development. Nontargeted techniques, including time-of-flight mass spectrophotometers (TOF-MS), may also be helpful. Rather than monitoring for individual chemicals, TOF-MS has been important for the progress of biomonitoring in recent years by allowing researchers to monitor for tens of thousands of organic compounds at a time. This enables researchers to circumvent policy issues that do not require companies to disclose the compounds they employ in their activities, such as is the case in many regions throughout the United States.

Even with full disclosure of the chemicals added to fracturing fluid, the ability to link chemicals to specific health outcomes remains difficult. Fracturing fluids and flowback and produced wastewaters are complex mixtures of chemicals with individual and possibly cumulative and synergistic properties. Many health outcomes are not specific to chemicals associated with shale gas development (e.g., headaches can be caused by a number of factors, rashes can be nonspecific, and asthma can be induced through a number of pathways), complicating the task of assessing associations between exposures and health outcomes. In turn, more exposure assessments and water and air monitoring should be undertaken to investigate the full suite of compounds emitted to the environment from these activities.

The chemicals contained in fracturing fluids are often not publicly disclosed because of trade secret laws and exemptions under the Energy Policy Act of 2005 that further confound environmental public health research. Moreover, the U.S. EPA is precluded from regulating hydraulic fracturing under the Safe Drinking Water Act (1974), and Congress expressly exempted hydraulic fracturing from the Underground Injection Control program (U.S. EPA 2012b). The nondisclosure of these chemicals creates research barriers because it is difficult to monitor for unknown compounds.

Limitations

In this review, we focused on the peer-reviewed scientific literature on the environmental public health dimensions of shale gas development. Although we used a broad search strategy, some publications and other relevant data could have been missed in our literature searches. However, we consider this to be a substantive summary of the currently available literature. Results of future studies will clarify the scientific understanding of the environmental public health concerns of shale gas development.

CONCLUSIONS

We reviewed the body of evidence of potential environmental public health dimensions of shale gas development. Scientific modeling and field investigations have helped to illuminate the emerging environmental issues with which shale gas production may be associated. Several studies have suggested that shale gas development contributes to pollutants in ambient air at concentrations known to be associated with increased risk of morbidity and mortality (Colborn et al. 2014; Kemball-Cook et al. 2010; McKenzie et al. 2012,2014). Similarly, some evidence supports theories of water contamination risks through a variety of pathways, most notably during wastewater transport and disposal and through failed cement in wells with poor structural integrity (Vengosh et al. 2013; Vidic et al. 2013;Warner et al. 2013a). The existing peer-reviewed scientific data suggest that there are potential risks that could possibly influence public health. More research is needed to clarify the magnitude of these concerns. Because shale gas development activities have accelerated dramatically over the past decade, the need for well-designed empirical studies becomes increasingly apparent.

REFERENCES

1. Agency for Toxic Substances and Disease Registry. 2005. Public Health Assessment Guidance Manual (Update). Available: http://www.atsdr.cdc.gov/hac/PHAManual/PDFs/PHAGM_final1-27-05.pdf [accessed 15 July 2013].
2. Alley B, Beebe A, Rodgers J Jr, Castle JW. 2011. Chemical and physical characterization of produced waters from conventional and unconventional fossil fuel resources. Chemosphere 85:74–82.
3. Aminto A, Olson MS. 2012. Four-compartment partition model of hazardous components in hydraulic fracturing fluid additives. J Nat Gas Sci Eng 7:16–21.
4. Balaba RS, Smart RB. 2012. Total arsenic and selenium analysis in Marcellus shale, high-salinity water, and hydrofracture flowback wastewater. Chemosphere 89:1437–1442.
5. Bamberger M, Oswald RE. 2012. Impacts of gas drilling on human and animal health. New Solut 22:51–77.
6. Bourgoyne AT Jr, Scott SL, Manowski W. 2000. A Review of Sustained Casing Pressure Occurring on the OCS. Available: http://www.wellintegrity.net/Documents/MMS%20-%20Review%20of%20SCP%20-%202000.pdf [accessed 15 August 2013].
7. Brufatto C, Cochran J, Conn L, Power D, El-Zeghaty SZ, Fraboulet B, et al. 2003. From Mud to Cement—Building Gas Wells. Available: http://www.slb.com/~/media/Files/resources/oilfield_review/ors03/aut03/p62_76.pdf [accessed 14 July 2013].
8. Bunch AG, Perry CS, Abraham L, Wikoff DS, Tachovsky JA, Hixon JG, et al. 2013. Evaluation of impact of shale gas operations in the Barnett Shale region on volatile organic compounds in air and

potential human health risks. Sci Total Environ 468–469:832–842.

9. Carter WPL, Seinfeld JH. 2012. Winter ozone formation and VOC incremental reactivities in the Upper Green River Basin of Wyoming. Atmos Environ 50:255–266.

10. CDC (Centers for Disease Control and Prevention). 1992. Preventing Silicosis and Deaths in Rock Drillers. NIOSH 92-107. Available: http://www.cdc.gov/niosh/docs/92-107/ [accessed 13 June 2013].

11. CDC (Centers for Disease Control and Prevention). 2002. Health Effects of Occupational Exposure to Respirable Crystalline Silica. NIOSH 2002-129. Available:http://www.cdc.gov/niosh/docs/2002-129/ [accessed 5 June 2013].

12. Centner TJ. 2013. Oversight of shale gas production in the United States and the disclosure of toxic substances. Resources Policy 38:233–240.

13. Centner TJ, O'Connell LK. 2014. Unfinished business in the regulation of shale gas production in the United States. Sci Total Environ 476–477:359–367.

14. Chilingar GV, Endres B. 2005. Environmental hazards posed by the Los Angeles Basin urban oilfields: an historical perspective of lessons learned. Environ Geol 47:302–317.

15. Clean Water Act of 1972. 1972. Public Law 92-500. Available:http://www.gpo.gov/fdsys/pkg/STATUTE-86/pdf/STATUTE-86-Pg816.pdf [accessed 2 July 2014].

16. Colborn T, Kwiatkowski C, Schultz K, Bachran M. 2011. Natural gas operations from a public health perspective. Hum Ecol Risk Assess 17:1039–1056.

17. Colborn T, Schultz K, Herrick L, Kwiatkowski C. 2014. An exploratory study of air quality near natural gas operations. Hum Ecol Risk Assess 20:86–105.

18. Davies RJ. 2011. Methane contamination of drinking water caused by hydraulic fracturing remains unproven. Proc Natl Acad Sci USA 108:E871; doi: 10.1073/pnas.1113299108.

19. Diamanti-Kandarakis E, Bourguignon JP, Giudice LC, Hauser R, Prins GS, Soto AM, et al. 2009. Endocrine-disrupting chemicals:

an Endocrine Society scientific statement. Endocr Rev 30:293–342.

20. DiGiulio DC, Wilken RT, Miller C, Oberley G. 2011. Draft: Investigation of Ground Water Contamination near Pavillion, Wyoming. Available:http://www2.epa.gov/sites/production/files/documents/EPA_ReportOnPavillion_Dec-8-2011.pdf [accessed 2 July 2014].
21. Edwards PM, Young CJ, Aikin K, deGouw JA, Dubé WP, Geiger F, et al. 2013. Ozone photochemistry in an oil and natural gas extraction region during winter: simulations of a snow-free season in the Uintah Basin, Utah. Atmos Chem Phys Discuss 13:7503–7552.
22. Energy Policy Act of 2005. 2005. Public Law 109-58. Available:http://www.gpo.gov/fdsys/pkg/PLAW-109publ58/html/PLAW-109publ58.htm [accessd 2 July 2014].
23. Esswein EJ, Breitenstein M, Snawder J, Kiefer M, Sieber WK. 2013. Occupational exposures to respirable crystalline silica during hydraulic fracturing. J Occup Environ Hyg 10:347–356.
24. Ferrar KJ, Michanowicz DR, Christen CL, Mulcahy N, Malone SL, Sharma RK. 2013. Assessment of effluent contaminants from three facilities discharging Marcellus Shale wastewater to surface waters in Pennsylvania. Environ Sci Technol 47:3472–3481.
25. Finkel M, Hays J, Law A. 2013. The shale gas boom and the need for rational policy. Am J Public Health 103(7):1161–1163.
26. Fontenot BE, Hunt LR, Hildenbrand ZL, Carlton DD, Oka H, Walton JL, et al. 2013. An evaluation of water quality in private drinking water wells near natural gas extraction sites in the Barnett Shale Formation. Environ Sci Technol 47(17):10032–10040.
27. Fryzek J, Pastula S, Jiang X, Garabrant DH. 2013. Childhood cancer incidence in Pennsylvania counties in relation to living in counties with hydraulic fracturing sites. J Occup Environ Med 55(7):796–801.
28. Garshick E, Laden F, Hart JE, Rosner B, Davis ME, Eisen EA, et al. 2008. Lung cancer and vehicle exhaust in trucking industry workers. Environ Health Perspect 116:1327–1332; doi:10.1289/ehp.11293.

29. Goldstein BD, Kriesky J, Pavliakova B. 2012. Missing from the table: role of the environmental public health community in governmental advisory commissions related to Marcellus Shale drilling. Environ Health Perspect 120:483–486; doi: 10.1289/ehp.1104594.
30. Gross SA, Avens HJ, Banducci AM, Sahmel J, Panko JM, Tvermoes BE. 2013. Analysis of BTEX groundwater concentrations from surface spills associated with hydraulic fracturing operations. J Air Waste Manag Assoc 63(4):424–432.
31. Haluszczak LO, Rose AW, Kump LR. 2013. Geochemical evaluation of flowback brine from Marcellus gas wells in Pennsylvania, USA. Appl Geochem 28:55–61.
32. Hladik ML, Focazio MJ, Engle M. 2014. Discharges of produced waters from oil and gas extraction via wastewater treatment plants are sources of disinfection by-products to receiving streams. Sci Total Environ 466–467:1085–1093.
33. Horn AD. 2009. Breakthrough Mobile Water Treatment Converts 75% of Fracturing Flowback Fluid to Fresh Water and Lowers CO_2 Emissions. Available:http://content.stockpr.com/esph2/files/techreports/SPE121101_Paper_March_2009.pdf[accessed 7 February 2014].
34. Hughes JD. 2013. Energy: a reality check on the shale revolution. Nature 494:307–308.
35. Jackson RB, Pearson BR, Osborn SG, Warner NR, Vengosh A. 2011. Research and Policy Recommendations for Hydraulic Fracturing and Shale-Gas Extraction. Available:http://www.nicholas.duke.edu/cgc/HydraulicFracturingWhitepaper2011.pdf [accessed 14 July 2013].
36. Jackson RB, Vengosh A, Darrah TH, Warner NR, Down A, Poreda RJ, et al. 2013. Increased stray gas abundance in a subset of drinking water wells near Marcellus shale gas extraction. Proc Natl Acad Sci USA 110(28):11250–11255; doi: 10.1073/pnas.1221635110.
37. Jerrett M, Burnett RT, Pope CA III, Ito K, Thurston G, Krewski D, et al. 2009. Long-term ozone exposure and mortality. N Engl J Med 360:1085–1095.

38. Karion A, Sweeney C, Pétron G, Frost G, Hardesty RM, Kofler J, et al. 2013. Methane emissions estimate from airborne measurements over a western United States natural gas field. Geophys Res Lett 40(16):4393–4397.

39. Kassotis CD, Tillitt DE, Davis JW, Hormann AM, Nagel SC. 2014. Estrogen and androgen receptor activities of hydraulic fracturing chemicals and surface and ground water in a drilling-dense region. Endocrinology 155:897–907; doi: 10.1210/en.2013-1697. [Online 16 December 2013].

40. Kemball-Cook S, Bar-Ilan A, Grant J, Parker L, Jung J, Santamaria W, et al. 2010. Ozone impacts of natural gas development in the Haynesville Shale. Environ Sci Technol 44(24):9357–9363.

41. Konschnik K, Holden M, Shasteen A. 2013. Legal Fractures in Chemical Disclosure Laws. Why the Voluntary Chemical Disclosure Registry FracFocus Fails as a Regulatory Compliance Tool. Available: http://blogs.law.harvard.edu/environmentallawprogram/files/2013/04/4-23-2013-LEGAL-FRACTURES.pdf [accessed 18 September 2013].

42. Korfmacher KS, Jones WA, Malone SL, Vinci LF. 2013. Public health and high volume hydraulic fracturing. New Solut 23:13–31.

43. Kortenkamp A, Faust M, Scholze M, Backhaus T. 2007. Low-level exposure to multiple chemicals: reason for human health concerns? Environ Health Perspect 115(suppl 1):106–114; doi: 10.1289/ehp.9358.

44. Leenheer JA, Noyes TI, Stuber HA. 1982. Determination of polar organic solutes in oil-shale retort water. Environ Sci Technol 16:714–723.

45. Legere L. 2013. DEP Drilling Records Reveal Water Damage. Available:http://citizensvoice.com/news/dep-drilling-records-reveal-water-damage-1.1491985[accessed 10 June 2013].

46. Lutz BD, Lewis AN, Doyle MW. 2013. Generation, transport, and disposal of wastewater associated with Marcellus Shale gas development. Water Resour Res 49:647–656.

47. Maule AL, Makey CM, Benson EB, Burrows IJ, Scammell MK. 2013. Disclosure of hydraulic fracturing fluid chemical additives: analysis of regulations. New Solut 23:167–187.

48. McKenzie LM, Guo R, Witter RZ, Savitz DA, Newman LS, Adgate JL. 2014. Birth outcomes and maternal residential proximity to natural gas development in rural Colorado. Environ Health Perspect 122:412–417; doi: 10.1289/ehp.1306722 [Online 28 January 2014].

49. McKenzie LM, Witter RZ, Newman LS, Adgate JL. 2012. Human health risk assessment of air emissions from development of unconventional natural gas resources. Sci Total Environ 424:79–87.

50. Meteorological Solutions Inc. 2011. Final Report: 2011. Upper Green River Ozone Study. Available: http://deq.state.wy.us/aqd/downloads/AirMonitor/Final_UGWOS_2011_Ozone_Study_Report_Text_and_Appendices.pdf [accessed 23 December 2013].

51. Miller SM, Wofsy SC, Michalak AM, Kort EA, Andrews AE, Biraud SC, et al. 2013. Anthropogenic emissions of methane in the United States. Proc Natl Acad Sci USA 110(50):20018–20022; doi: 10.1073/pnas.1314392110 [Online 25 November 2013].

52. Myers T. 2012. Potential contaminant pathways from hydraulically fractured shale to aquifers. Ground Water 50:872–882.

53. New York State Department of Environmental Conservation. 2011. Revised Draft SGEIS on the Oil, Gas and Solution Mining Regulatory Program. Available:http://www.dec.ny.gov/energy/75370.html [accessed 14 June 2013].

54. Ohio Department of Natural Resources. 2008. Report on the Investigation of the Natural Gas Invastion of Aquifers in Bainbridge Township of Geauga County, Ohio. Available:http://s3.amazonaws.com/propublica/assets/natural_gas/ohio_methane_report_080901.pdf [accessed 19 August 2013].

55. Olaguer EP. 2012. The potential near-source ozone impacts of upstream oil and gas industry emissions. J Air Waste Manag Assoc 62:966–977.

56. Osborn SG, Vengosh A, Warner NR, Jackson RB. 2011. Methane contamination of drinking water accompanying gas-well drilling and hydraulic fracturing. Proc Natl Acad Sci USA 108:8172–8176; doi: 10.1073/pnas.1100682108.

57. Pennsylvania Department of Environmental Protection. 2010. PA DEP Oil & Gas Reporting Website. Available: https://www.paoilandgasreporting.state.pa.us/publicreports/Modules/Welcome/Agreement.aspx [accessed 18 September 2013].
58. Papoulias DM, Velasco AL. 2013. Histopathological analysis of fish from Acorn Fork Creek, Kentucky, exposed to hydraulic fracturing fluid releases. Southeast Nat 12:92–111.
59. Peischl J, Ryerson TB, Brioude J, Aikin KC, Andrews AE, Atlas E, et al. 2013. Quantifying sources of methane using light alkanes in the Los Angeles basin, California. J Geophys Res Atmos 118:4974–4990; doi: 10.1002/jgrd.50413.
60. Pétron G, Frost G, Miller BR, Hirsch AI, Montzka SA, Karion A, et al. 2012. Hydrocarbon emissions characterization in the Colorado Front Range: a pilot study. J Geophys Res Atmos 117(D4):D04303; doi: 10.1029/2011JD016360.
61. Pinto J. 2009. Atmospheric chemistry: Wyoming winter smog. Nature Geosci 2:88–89.
62. Pope CA III, Burnett RT, Thun MJ, Calle EE, Krewski D, Ito K, et al. 2002. Lung cancer, cardiopulmonary mortality, and long-term exposure to fine particulate air pollution. JAMA 287:1132–1141.
63. Pope CA III, Burnett RT, Thurston GD, Thun MJ, Calle EE, Krewski D, et al. 2004. Cardiovascular mortality and long-term exposure to particulate air pollution: epidemiological evidence of general pathophysiological pathways of disease. Circulation 109:71–77.
64. Rich AL, Crosby EC. 2013. Analysis of reserve pit sludge from unconventional natural gas hydraulic fracturing and drilling operations for the presence of technologically enhanced naturally occurring radioactive material (TENORM). New Solut 23:117–135.
65. Roy AA, Adams PJ, Robinson AL. 2013. Air pollutant emissions from the development, production, and processing of Marcellus Shale natural gas. J Air Waste Manag Assoc 64(1):19–37.
66. Rozell DJ, Reaven SJ. 2012. Water pollution risk associated with natural gas extraction from the Marcellus Shale. Risk Anal 32:1382–1393.
67. Saba T, Orzechowski M. 2011. Lack of data to support a relationship between methane contamination of drinking water

wells and hydraulic fracturing [Letter]. Proc Natl Acad Sci USA 108:E663; doi: 10.1073/pnas.1108435108.

68. Saberi P. 2013. Navigating medical issues in shale territory. New Solut 23:209–221.

69. Safe Drinking Water Act of 1974. 1974. Public Law 93-523. Available:http://www.gpo.gov/fdsys/pkg/STATUTE-88/pdf/STATUTE-88-Pg1660-2.pdf [accessed 2 July 2014].

70. Schnell RC, Oltmans SJ, Neely RR, Endres MS, Molenar JV, White AB. 2009. Rapid photochemical production of ozone at high concentrations in a rural site during winter [Letter]. Nat Geosci 2:120–122.

71. Schon SC. 2011. Hydraulic fracturing not responsible for methane migration [Letter]. Proc Natl Acad Sci USA 108:E664; doi: 10.1073/pnas.1107960108.

72. Smith KR, Jerrett M, Anderson HR, Burnett RT, Stone V, Derwent R, et al. 2009. Public health benefits of strategies to reduce greenhouse-gas emissions: health implications of short-lived greenhouse pollutants. Lancet 374:2091–2103.

73. Southwest Energy. 2012. Frac Fluid—What's in it? Available:http://www.swn.com/operations/documents/frac_fluid_fact_sheet.pdf [accessed 17 August 2013].

74. Steinzor N, Subra W, Sumi L. 2013. Investigating links between shale gas development and health impacts through a community survey project in Pennsylvania. New Solut 23:55–83.

75. Teuschler LK, Hertzberg RC. 1995. Current and future risk assessment guidelines, policy, and methods development for chemical mixtures. Toxicology 105(2–3):137–144.

76. Tran HT, Alvarado A, Garcia C, Motallebi N, Miyasato L, Vance W. 2008. Methodology for Estimating Premature Deaths Associated with Long-term Exposure to Fine Airborne Particulate Matter in California. Sacramento, CA:California Air Resources Board. Available:http://www.arb.ca.gov/research/health/pm-mort/pm-mortdraft.pdf [accessed 16 August 2013].

77. United Nations Environment Programme. 2011. Integrated Assessment of Black Carbon and Tropospheric Ozone: Summary for Decision Makers. Available:http://www.unep.org/dewa/Portals/67/pdf/Black_Carbon.pdf [accessed 7 August 2013].

78. U.S. Department of Energy. 2011. The SEAB Shale Gas Production Subcommittee Ninety-Day Report—August 11, 2011. Available: http://www.shalegas.energy.gov/resources/081111_90_day_report.pdf [accessed 17 September 2013].
79. U.S. Energy Information Administration. 2013. How Much Carbon Dioxide Is Produced When Different Fuels Are Burned? Available: http://www.eia.gov/tools/faqs/faq.cfm?id=73&t=11 [accessed 17 September 2013].
80. U.S. EPA (U.S. Environmental Protection Agency). 2010a. Hydraulic Fracturing Research Study. Available: http://www.epa.gov/safewater/uic/pdfs/hfresearchstudyfs.pdf [accessed 14 June 2013].
81. U.S. EPA (U.S. Environmental Protection Agency). 2010b. Range Resources Imminent and Substantial Endangerment Order, Parker County, TX. Available:http://www.epa.gov/region6/region-6/tx/tx005.html [accessed 13 August 2013].
82. U.S. EPA (U.S. Environmental Protection Agency). 2011a. EPA Releases Draft Findings of Pavillion, Wyoming Ground Water Investigation for Public Comment and Independent Scientific Review [Press Release]. Available: http://yosemite.epa.gov/opa/admpress.nsf/20ed1dfa1751192c8525735900400c30/ef35bd26a80d6ce3852579600065c94e!OpenDocument [accessed 19 June 2013].
83. U.S. EPA (U.S. Environmental Protection Agency). 2011b. Plan to Study the Potential Impacts of Hydraulic Fracturing on Drinking Water Resources. Available:http://water.epa.gov/type/groundwater/uic/class2/hydraulicfracturing/upload/hf_study_plan_110211_final_508.pdf [accessed 15 August 2013].
84. U.S. EPA (U.S. Environmental Protection Agency). 2012a. National Ambient Air Quality Standards (NAAQS). Available: http://www.epa.gov/air/criteria.html [accessed 1 July 2014].
85. U.S. EPA (U.S. Environmental Protection Agency). 2012b. Class II Wells – Oil and Gas Related Injection Wells (Class II). Available: http://water.epa.gov/type/groundwater/uic/class2/index.cfm [accessed 16 July 2014].
86. U.S. EPA (U.S. Environmental Protection Agency). 2013. Ozone. Available:http://www.epa.gov/ncea/isa/ozone.htm [accessed 7 August 2013].

87. U.S. House of Representatives, Committee on Energy and Commerce. 2011. Chemicals Used in Hydraulic Fracturing. Available: http://democrats.energycommerce.house.gov/sites/default/files/documents/Hydraulic-Fracturing-Chemicals-2011-4-18.pdf [accessed 17 December 2013].

88. Utah Department of Environmental Quality. 2013. Final Report: 2012 Uintah Basin Winter Ozone & Air Quality Study. Available: http://rd.usu.edu/files/uploads/ubos_2011-12_final_report.pdf [accessed 19 September 2013].

89. Vandenberg LN, Colborn T, Hayes TB, Heindel JJ, Jacobs DR Jr, Lee DH, et al. 2012. Hormones and endocrine-disrupting chemicals: low-dose effects and nonmonotonic dose responses. Endocr Rev 33:378–455.

90. Vengosh A, Warner N, Jackson R, Darrah T. 2013. The effects of shale gas exploration and hydraulic fracturing on the quality of water resources in the United States. Procedia Earth Planetary Sci 7:863–866.

91. Vidic RD, Brantley SL, Vandenbossche JM, Yoxtheimer D, Abad JD. 2013. Impact of shale gas development on regional water quality. Science 340 (6134); doi: 10.1126/science.1235009.

92. Warner NR, Christie CA, Jackson RB, Vengosh A. 2013a. Impacts of shale gas wastewater disposal on water quality in western Pennsylvania. Environ Sci Technol 47(20):11849–11857.

93. Warner NR, Jackson RB, Darrah TH, Osborn SG, Down A, Zhao K, et al. 2012. Geochemical evidence for possible natural migration of Marcellus Formation brine to shallow aquifers in Pennsylvania. Proc Natl Acad Sci USA 109:11961–11966; doi: 10.1073/pnas.1121181109.

94. Warner NR, Kresse TM, Hays PD, Down A, Karr JD, Jackson RB, et al. 2013b. Geochemical and isotopic variations in shallow groundwater in areas of the Fayetteville Shale development, north-central Arkansas. Appl Geochem 35:207–220.

95. Watson T, Bachu S. 2009. Evaluation of the potential for gas and CO_2 leakage along wellbores. SPE Drilling Completion 24:115–126.

96. Weinhold B. 2008. Ozone nation: EPA standard panned by the people. Environ Health Perspect 116:A302–A305.

97. Wilkinson CF, Christoph GR, Julien E, Kelley JM, Kronenberg J, McCarthy J, et al. 2000. Assessing the risks of exposures to multiple chemicals with a common mechanism of toxicity: how to cumulate? Regul Toxicol Pharmacol 31(1):30–43.

98. Wilson JM, VanBriesen JM. 2012. Oil and gas produced water management and surface drinking water sources in Pennsylvania. Environ Pract 14:288–300.

99. Witter RZ, McKenzie L, Stinson KE, Scott K, Newman LS, Adgate J. 2013. The use of health impact assessment for a community undergoing natural gas development. Am J Public Health 103:1002–1010.

100. Zoeller RT, Brown TR, Doan LL, Gore AC, Skakkebaek NE, Soto AM, et al. 2012. Endocrine-disrupting chemicals and public health protection: a statement of principles from The Endocrine Society. Endocrinology 153:4097–4110.

Chapter 6

Synthesis of Copper Micro-Rods with Layered Nano-Structure by Thermal Decomposition of the Coordination Complex Cu(BTA)2

Botao Qu, Xinrong Lu, Yan Wu, Xiaozeng You, and Xiangxing Xu

State Key Laboratory of Coordination Chemistry, School of Chemistry and Chemical Engineering, Collaborative Innovation Center of Advanced Microstructures, Nanjing University, Nanjing 210093, People's Republic of China

ABSTRACT

Porous metallic copper was successfully prepared by a simple thermal decomposition strategy. A coordination compound of Cu $(BTA)_2$ with the morphology of micro-rod crystal was synthesized as the precursor. The precursor to copper transformation was performed and annealed at 600°C with the shape preserved. The copper micro-rods are assembled from unique thin lamellar layers, each with the thickness of approximately 200 nm and nano-pores of approximately 20 to 100 nm. This morphology is highly related to the crystal structure of the precursor. The mechanism of the morphology formation is proposed, which would be able to offer a guideline toward porous metals with controllable macro/micro/nano-structures by the precursor crystal growth and design.

BACKGROUND

Porous metallic materials have become a burgeoning field in both applied technology and basic scientific research, especially for their significant thermal or electron conductivity, catalysis properties, importance in interface engineering, energy industry, and biomedical applications [1]-[4]. During the past two decades, the synthesis methods of porous metals evolved with the development of nano-science and nano-technology. Sol-gel, dealloying, and soft-template methods are typical synthetic strategies [5]-[8]. Among the periodic table of elements, metals of silver, nickel, copper, palladium, ruthenium, titanium, and platinum have been intensively investigated for their porous foam. For example, Walsh et al. used dextran as a sacrificial template to fabricate silver sponges, Yamauchi and co-workers prepared mesoporous nickel by an electroless deposition method in the presence of lyotropic liquid crystals, and Kuroda et al. reported 2D hexagonally ordered mesoporous metals (Ru, Pt, or Pd) by dissolving silica replica [9]-[11]. In addition to these methods, the combustion technology is a general method to synthesize various porous metals and oxides [12]-[14]. Compared with porous noble metals [15]-[17], porous copper was less focused, possibly for its high reactivity of oxidization in ambient atmosphere [18]-[26].

In this report, porous metallic copper micro-rods (10 to 1,000 μm) with layered nano-structure were successfully synthesized by a thermal decomposition method. Figure 1 shows the proposed synthesis diagram. The coordination compound of $Cu(BTA)_2$ (BTA: bis[1(2) H-tetrazol-5-yl]amine) as the precursor was synthesized, based on the following merits. First, among various coordination compounds that contain high-nitrogen ligands, $Cu(BTA)_2$ is mildly energetic during combustion process, which is favored for both the reaction and the porous structure formation [27],[28]. Too high energetic precursors would destroy the porous micro/nano-structure, while less energetic precursors would lead to incomplete decomposition that induces mass impurities. Second, the absence of oxide in BTA prevents the products from oxidization. Nitrogen and carbon can be largely eliminated in the gas form from the material during the combustion process. Third, the T_d point group molecule configuration and crystalline morphology of the $Cu(BTA)_2$ may contribute to the micro/nano-structure of the metallic face-centered cubic (fcc) copper. Therefore, the micro/nano-structure could be rationally tuned and tailored by means of a precursor molecule and corresponding crystal growth and design. Finally, in comparison with metallurgy methods, the decomposition is performed at a relatively low temperature with properties of low density and gas permeability originating from the porous feature.

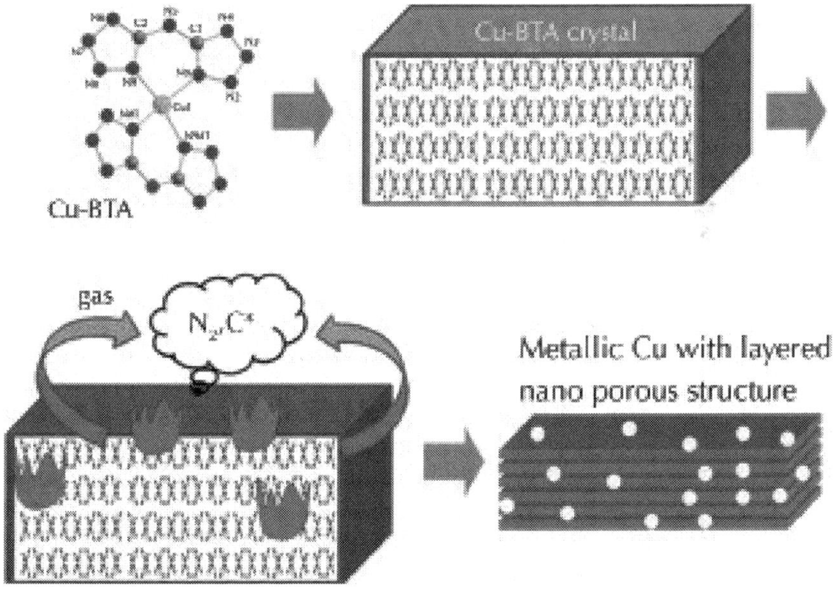

Figure 1: Schematic diagram of the decomposition of a Cu(BTA)2crystal to form porous copper.

METHODS

Materials and Equipment

All chemicals were purchased from commercial sources (Sigma-Aldrich, St. Louis, MO, USA) and were used without further purification. Single-crystal X-ray diffraction measurements were carried out on a Bruker SMART APEX CCD (Bruker AXS, Inc., Madison, WI, USA). Thermogravimetric analyses (TGA) were measured on a simultaneous SDT 2960 thermal analyzer with a heating rate of 20°C·min^{-1}under N_2 atmosphere. Powder X-ray diffraction (XRD) patterns were collected using a Bruker D8 ADVANCE X-ray diffractometer equipped with Cu-Kα radiation (λ=1.5418 Å) at 40 kV and 40 mA. Scanning electron microscopy (SEM) characterizations were performed on a Hitachi S-4800 SEM (Hitachi, Ltd, Chiyoda-ku, Japan), equipped with energy-dispersive X-ray spectroscopy (EDS). The transmission electron microscopy (TEM) images were obtained from JEOL JEM-2100 (JEOL Ltd., Akishima-shi, Japan) operating at 200 kV. The electronic semiconducting property f the samples was recorded by the semiconductor device analyzer B1500A from Agilent Technologies (Santa Clara, CA, USA). Elemental analysis is measured by Elementar vario MICRO (Hanau, Germany). The Fourier transform infrared (FTIR) spectrum was measured by a VECTOR 22 spectrometer with KBr pellets (Bruker AXS, Inc., Madison, WI, USA).

Synthesis

A mixture of $CuCl_2 \cdot 2H_2O$ (AR, 0.1 mmol), BTA (99%, 0.1 mmol), and $NH_3 \cdot H_2O$ (AR, 25 to 28 wt%, 1.0 ml) in CH_3CN (AR, 5 ml) was sealed in a Teflon-lined stainless steel autoclave and heated to 100°C under autogenous pressure for 48 h. When cooled to room temperature, blue rodlike $Cu(BTA)_2$crystals were isolated. They were rinsed with CH_3CN and dried in vacuum at 60°C overnight. The $Cu(BTA)_2$ was decomposed and annealed in an Ar flow atmosphere at 600°C to yield the porous metallic copper.

RESULTS AND DISCUSSION

Single-crystal X-ray diffraction reveals that the crystal structure of the precursor belongs to the monoclinic space group *C2/c* (Figure 1). The mononuclear complex contains one four-coordinated Cu(II) cation and two BTA anions with the formula $Cu(BTA)_2$. Each of the BTA ligands provides two donor N atoms to coordinate to the Cu(II) ions, exhibiting a distorted tetrahedron coordination geometry (Additional file 1: Figure S1). The Cu-N bond lengths for each ligand are 1.954 and 1.969 Å, and the dihedral angle of two BTA planes is 36.35°. The CCDC no. 1035211 contains the supplementary crystallographic data for the $Cu(BTA)_2$ complex (Additional file 2).

The TGA curve of the $Cu(BTA)_2$ complex shows that there exist clear stages for the thermal decomposition. The material remains stable until the temperature rises to 240°C that the compound becomes decomposed. It undergoes a steep weight loss at 325°C and ends at approximately 340°C (Figure 2a). A minor weight loss was observed within 400°C to 550°C, which could be due to the further decomposition of carbon by-products [28]. It means that the precursor can be ignited at a temperature as low as 325°C to 340°C. The annealing temperature of 600°C was used to achieve higher product purity. During the self-sustained combustion, the Cu^{2+} ions are reduced to metal Cu^0atoms. These atoms agglomerate into small grains, which assemble into large structures along with the generation of N_2- and carbon-containing gases.

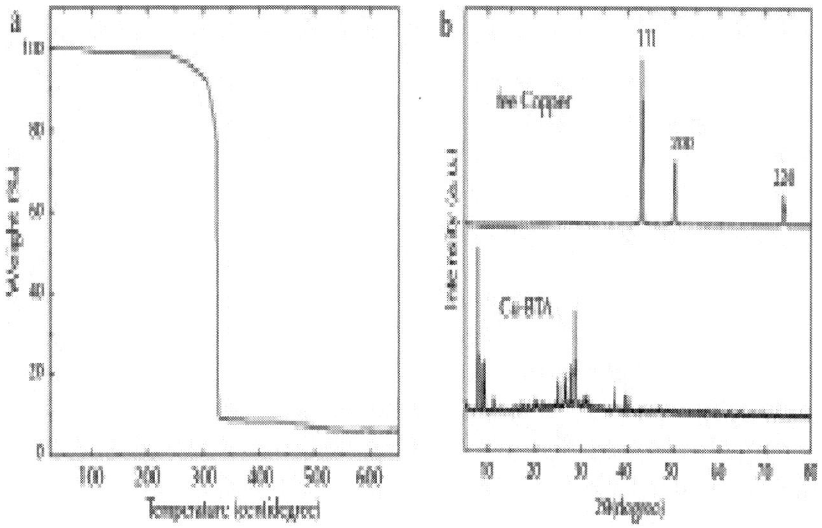

Figure 2: TGA curve and XRD patterns. (a) TGA curve of the Cu(BTA)2 precursor. (b) XRD patterns of the $Cu(BTA)_2$ crystals and the copper product.

The XRD measurement reveals that after the decomposition at 600°C, $Cu(BTA)_2$ completely transformed into metallic copper. The diffraction peaks (Figure 2b) can be indexed to (111), (200), and (220) crystal planes of fcc copper (pdf #88-1326). The original morphology of the $Cu(BTA)_2$crystals was largely inherited by the metallic copper, showing a short rodlike shape, while at the same time, remarkable size shrinkage (up to approximately 90% volume shrinkage) is observed for the metallic copper compared with the $Cu(BTA)_2$ crystal precursor.

The porous feature is confirmed by the SEM measurement. Typical SEM images are shown in Figure 3a,b. A unique nano-structure was observed for the sample. All these copper micro-rods are assembled by loosely piled nano-sheets with the thickness approximately 200 nm. The gaps between the nano-sheet layers are approximately 0.5 to 2 μm. Nano-pores of the size approximately 20 to 100 nm are also observed within the nano-sheets. The porous structures - both the micro-gaps and nano-pores - are believed to be induced by the N_2- and carbon-containing gases generated during the decomposition. Interestingly, the piled copper nano-sheets are well arranged parallel to the axial direction. The EDS spectrum showed that the copper product is mainly composed of copper and some minor carbon (Figure 3c). The minor carbon should be from the residual by-products in the decomposition procedure. Elementary analysis shows that the C content is 0.59 wt%. The TEM measurement confirms that a thin layer (approximately 4 nm) of amorphous carbon exists in the outer layer of the copper (Figure 3d). It is still unclear if or to what extent the carbon exists in the copper crystal in the form of an interstitial solid solution.

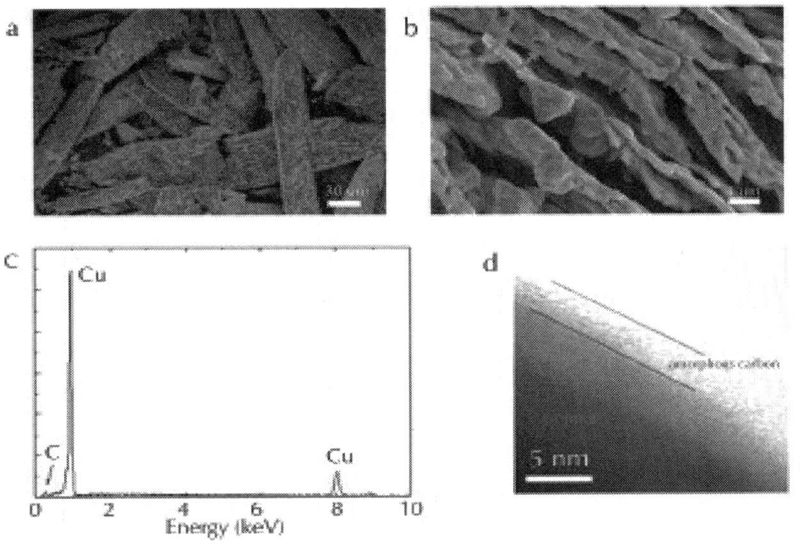

Figure 3: SEM images, EDS spectrum, and TEM image. (a, b) SEM images of the porous copper layered structure. Red arrows: small pores; blue arrows: gaps between layers. (c) EDS spectrum. (d) TEM image of the copper sample.

To understand how this orientation is formed, we take a closer look at the $Cu(BTA)_2$ crystal structure (Figure 4). The $Cu(BTA)_2$ micro-rods grow in the [010] direction. In the (100) crystal plane, the adjacent Cu-Cu distances are, respectively, 5.9, 6.2, and 7.0 Å, while in the [100] direction, the Cu-Cu distance is as large as 10.2 Å. Therefore, the $Cu(BTA)_2$ presents a layered crystal structure. This layered structure matches well with the layered structure of the corresponding copper. From this clue, the mechanism of the copper micro/nano-structure formation is suggested. The decomposition of each $Cu(BTA)_2$ molecule generates a single Cu^0 atom at the position *in situ*. These isolated Cu^0 atoms are highly active to attract each other forming a Cu-Cu metallic bond. The Cu-Cu metallic bond has the length of 3.6 Å, smaller than the Cu-Cu distance in the precursor. So the Cu^0 atoms would preferably approach from the short distance in a statistical thermodynamic way, that is, in the (100) crystal plane of the precursor. Thus, it intends to form a Cu layer structure with the thickness in one-atom scale. However, as the Cu-Cu bond formed, it obeys the lowest energy rule to transform into the fcc phase instantly. These initially formed Cu nano-clusters undergo alloying to form bigger crystals. The degassing process during the decomposition may contribute to the layered nano-structure: the decomposed N_2- and carbon-containing gases would escape preferably through the gaps between the layers, which may generate oriented pressure to shape the metallic copper.

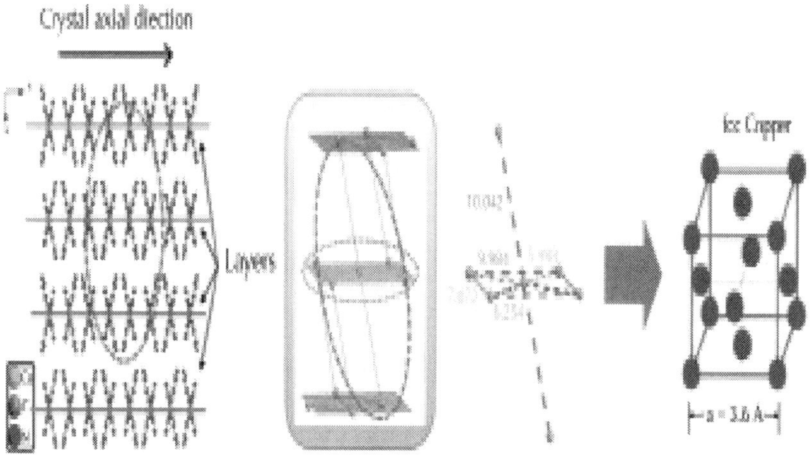

Figure 4: Schematic illustration of the Cu(BTA)2crystal structure and copper transformation.

The electrical conductivity of one Cu micro-rod was measured at room temperature (Figure 5). The electric conductivity is approximately 10^5 S·m^{-1} in the longitudinal direction. It is higher than amorphous carbon (approximately 2×10^4 to 3×10^4 S·m^{-1}) and lower than pure copper (5.96×10^7S·m^{-1}) [29], which is consistent with the theory of the electrical conductivity of composites [30]. In comparison, the copper micro-rod sample was further pressed into a rectangular block with the size of $0.888 \times 4 \times 12$ mm under 8 Mpa for 5 min. The electrical conductivity was found to be 7.4×10^6S·m^{-1}, which is one order of magnitude higher than that of the Cu micro-rod. It indicates that the electrical property depends not only on the component fractions, size, and shape but also on the porous structure: when intensively pressed, the pores are eliminated, leading to the enhancement of the electrical conductivity [30]. The micro/nano-porous structure and good conductivity suggest that the material may have application potentials in batteries/cells [31]-[33], sensing devices [34], and catalysis [35].

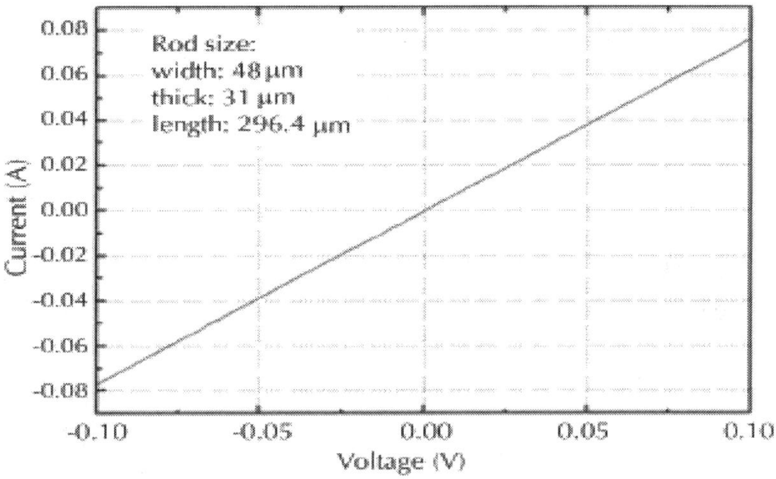

Figure 5: I-Vmeasurement of a copper micro-rod along the longitudinal direction.

CONCLUSIONS

In summary, we report a thermal decomposition method to prepare copper micro-rods with layered porous structure for the first time, by using a well-designed coordination compound of $Cu(BTA)_2$crystal as the precursor. The shape of the crystals was preserved for the copper product. This allows us to obtain copper with various morphologies by the growth of the precursor crystals of different sizes and shapes, without changing the molecule itself. Moreover, the layered nano-structure is highly related to the crystal parameters of the precursor. It could be expected that the same precursor crystallizes in different crystalline spaces; accordingly, the micro/nano-structure would be tuned.

AUTHORS' CONTRIBUTIONS

BTQ and XRL contributed equally to the manuscript. XZY and XXX designed the research. BTQ, XRL, YW, and XXX carried out the experiments and drafted the manuscript. All authors read and approved the final manuscript.

ACKNOWLEDGEMENTS

This work was supported by the Major State Basic Research Development Program of China (Grant Nos. 2013CB922102 and 2011CB808704), the National Natural Science Foundation of China (Grant Nos. 91022031 and 21301089), Jiangsu Province Science Foundation for Youths (BK20130562), and the Natural Science Foundation of Jiangsu Province (BK20130054).

REFERENCES

1. Li Y, Fu ZY, Su BL: Hierarchically structured porous materials for energy conversion and storage. *Adv Funct Mater.* 2012, 22:4634-67
2. Valtchev V, Tosheva L: Porous nanosized particles: preparation, properties, and applications. *Chem Rev.* 2013, 113:6734-60
3. Zhang J, Li CM: Nanoporous metals: fabrication strategies and advanced electrochemical applications in catalysis, sensing and energy systems. *Chem Soc Rev.* 2012, 41:7016-31
4. Brockway L, Vasiraju V, Asayesh-Ardakani H, Shahbazian-Yassar R, Vaddiraju S:Thermoelectric properties of large-scale Zn_3P_2 nanowire assemblies. *Nanotechnol.* 2014, 25:145401-8
5. Warren SC, Perkins MR, Adams AM, Kamperman M, Burns AA, Arora H, *et al.*: A silica sol–gel design strategy for nanostructured metallic materials. *Nat Mater.* 2012, 11:460-7
6. Chen LY, Yu JS, Fujita T, Chen MW: Nanoporous copper with tunable nanoporosity for SERS applications. *Adv Funct Mater.* 2009, 19:1221-6

7. Yamauchi Y, Kuroda K: Rational design of mesoporous metals and related nanomaterials by a soft-template approach. *Chem Asia J.* 2008, 3:664-76
8. Warren SC, Messina LC, Slaughter LS, Kamperman M, Zhou Q, Gruner SM, *et al.*: Ordered mesoporous materials from metal nanoparticle-block copolymer self-assembly. *Science.* 2008, 320:1748-52
9. Walsh D, Arcelli L, Ikoma T, Tanaka J, Mann S: Dextran templating for the synthesis of metallic and metal oxide sponges. *Nat Mater.* 2003, 2:386-90
10. Yamauchi Y, Yokoshima T, Mukaibo H, Tezuka M, Shigeno T, Momma T, *et al.*: Highly ordered mesoporous Ni particles prepared by electroless deposition from lyotropic liquid crystals. *Chem Lett.* 2004, 33:542-3
11. Takai A, Doi Y, Yamauchi Y, Kuroda K: A rational repeating template method for synthesis of 2D hexagonally ordered mesoporous precious metals. *Chem Asia J.* 2011, 6:881-7
12. Guo L, Arafune H, Teramae N: Synthesis of mesoporous metal oxide by the thermal decomposition of oxalate precursor. *Langmuir.* 2013, 29:4404-12
13. Zhang L, Wu HB, Xu R, Lou XW: Porous Fe_2O_3 nanocubes derived from MOFs for highly reversible lithium storage. *CrystEngComm.* 2013, 15:9332-5
14. Banerjee A, Gokhale R, Bhatnagar S, Jog J, Bhardwaj M, Lefez B, *et al.*: MOF derived porous carbon–Fe_3O_4 nanocomposite as a high performance, recyclable environmental superadsorbent. *J Mater Chem.* 2012, 22:19694-9
15. Lee MN, Mohraz A: Hierarchically porous silver monoliths from colloidal bicontinuous interfacially jammed emulsion gels. *J Am Chem Soc.* 2011, 133:6945-7
16. Qi J, Motwani P, Gheewala M, Brennan C, Wolfe JC, Shih WC: Surface-enhanced Raman spectroscopy with monolithic nanoporous gold disk substrates. *Nanoscale.* 2013, 5:4105-9
17. Lu L, Eychmüller A: Ordered macroporous bimetallic nanostructures: design, characterization, and applications. *Acc Chem Res.* 2008, 41:244-53

18. Shin HC, Liu M: Copper foam structures with highly porous nanostructured walls. *Chem Mater.* 2004, 16:5460-4
19. Kranzlin N, Niederberger M: Wet-chemical preparation of copper foam monoliths with tunable densities and complex macroscopic shapes. *Adv Mater.* 2013, 25:5599-604
20. Lai M, Kulak AN, Law D, Zhang Z, Meldrum FC, Riley DJ: Profiting from nature: macroporous copper with superior mechanical properties. *Chem Commun.* 2007, 34:3547-3549
21. Zang D, Wu C, Zhu R, Zhang W, Yu X, Zhang Y: Porous copper surfaces with improved superhydrophobicity under oil and their application in oil separation and capture from water. *Chem Commun.* 2013, 49:8410-2
22. Liu Y, Chu Y, Zhuo Y, Dong L, Li L, Li M: Controlled synthesis of various hollow Cu nano/microstructures via a novel reduction route. *Adv Funct Mater.* 2007, 17:933-8
23. Liu Z, Yang Y, Liang J, Hu Z, Li S, Peng S, *et al.*: Synthesis of copper nanowires via a complex-surfactant-assisted hydrothermal reduction process. *J Phys Chem B.* 2003, 107:12658-61
24. Chang Y, Lye ML, Zeng HC: Large-scale synthesis of high-quality ultralong copper nanowires. *Langmuir.* 2005, 21:3746-8
25. Adner D, Korb M, Schulze S, Hietschold M, Lang H: A straightforward approach to oxide-free copper nanoparticles by thermal decomposition of a copper(I) precursor. *Chem Commun.* 2013, 49:6855-7
26. Tappan BC, Huynh MH, Hiskey MA, Chavez DE, Luther EP, Mang JT, *et al.*: Ultralow-density nanostructured metal foams: combustion synthesis, morphology, and composition. *J Am Chem Soc.* 2006, 128:6589-94
27. Friedrich M, Gálvez-Ruiz JC, Klapötke TM, Mayer P, Weber B, Weigand JJ: BTA copper complexes. *Inorg Chem.* 2005, 44:8044-52
28. Tappan BC, Steiner SA 3rd, Luther EP: Nanoporous metal foams. *Angew Chem.* 2010, 49:4544-65
29. Beaty HW, Fink DG: *Standard handbook for electrical engineers. 11th.* McGraw-Hill, New York; 1978.
30. Kerner EH: The electrical conductivity of composite media. *Proc Phys Soc B.* 1956, 69:802-7

31. Amiri O, Salavati-Niasari M, Sabet M, Ghanbari D: Synthesis and characterization of $CuInS_2$ microsphere under controlled reaction conditions and its application in low-cost solar cells. *Mater Sci Semicond Process*. 2013, 16:1485-94

32. Amiri O, Salavati-Niasari M, Rafiei A, Farangi M: 147% improved efficiency of dye synthesized solar cells by using CdS QDs, Au nanorods and Au nanoparticles.*RSC Adv*. 2014, 4:62356-61

33. Liu S, Sun SH, You XZ: Inorganic nanostructured materials for high performance electrochemical supercapacitors.*Nanoscale*. 2014, 6:2037-45

34. Kevin M, Ong WL, Lee GH, Ho GW: Formation of hybrid structures: copper oxide nanocrystals templated on ultralong copper nanowires for open network sensing at room temperature. *Nanotechnol*. 2011, 22:235701-10

35. Sun QC, Ding YC, Goodman SM, Funke HH, Nagpal P: Copper plasmonics and catalysis: role of electron–phonon interactions in dephasing localized surface plasmons. *Nanoscale*. 2014, 6:12450-7.

Chapter 7

Application of Membrane in Gas Separation Processes: its Suitability and Mechanisms

Reza Abedini and Amir Nezhadmoghadam

Department of Chemical Engineering, Faculty of Engineering, Ferdowsi University of Mashhad, Mashhad, Iran

ABSTRACT

In separation of gases, membranes offer the greatest potential. In interaction with a membrane, a high degree of permeability coupled with a large selectivity of a specific gaseous species ensures superior performances in the gas processing industry. Economically, the advantage of membrane separation technology is immense. The separation of natural gas by thin barriers termed as membranes is a dynamic and rapidly growing field, and it has been proven to be technically and economically superior to other emerging technologies. This superiority is due to certain advantages which membrane technology benefits from, including low capital investment, low weight, and space requirement and high process flexibility. In the past ten years, the membrane gas separation technology has advanced greatly and can now be regarded as a competitive industrial gas separation method. The benefits of "Membrane Technology" also include higher recovery of the desired gaseous effluent that can be reused for multiple purposes. This article discusses various experimental and analytical models related to current membrane technologies.

INTRODUCTION

Solution gases are gaseous hydrocarbons that exist under increased pressure in oil reservoirs. They are "in solution" with the liquid hydrocarbon phase in equilibrium to a separate vapor phase. These gases are mostly conserved, but approximately 8% of the solution gases are considered uneconomical to be stored and are routinely flared. Gas flaring in crude oil production is a malpractice as substantial volumes of carbon dioxide are emitted during the process. Flaring has impacts on global warming, ozone depletion, and acidification. It also leads to resource depletion due to the combustion of the materials used in the flare construction. Flaring gases to the atmosphere can cause harmful side effects to the environment [1]. They include:

Carbon Dioxide and Methane Emission

The main gaseous emissions responsible for "Global Warming" are carbon dioxide and methane. Methane gas possesses 35 times global warming potential than carbon dioxide. The only feasible method for reduction of the volume of the wasted gases without constraining oil production is to improve electrical power generation efficiency and minimize flaring by exploiting gas conservation opportunities.

Hydrogen Sulfide/Sulfur Dioxide Emissions

Sulfur dioxide does not contribute to the greenhouse effect. The main environmental concern with sulfur dioxide (SO_2) is that it is a source of acid rain. Hydrogen sulfide (H_2S) is a toxic and corrosive gas, which is rapidly oxidized to sulfur dioxide in the atmosphere.

NO_x Emissions in the Presence of Volatile Organic Compounds

The nitrogen oxides can lead to the formation of ozone and other powerful respiratory irritants. Ozone, in addition to its impact on human health, can significantly reduce plant growth. Nitrogen oxides also contribute to the formation of acid rain and may contribute to global warming. Flaring is prevented only when the impurities in the gases are removed before transportation to the pipeline. The components to be removed from the flare gas include hydrogen sulfide, sulfur dioxide, carbon dioxide, and nitrogen. In the last few years, many efforts have been made to develop effective means to separate the impurities in natural gases. Ongoing research is being conducted to develop new membrane materials and improve pressure swing adsorption techniques to separate the gaseous mixtures. Membrane separation processes has become one of the emerging technologies, which have undergone a rapid growth during the past few decades. A membrane is defined as a selective barrier between two fluid phases [2-4]. Gas separation became a major industrial application of membrane technology only during the past 15 years, but the study of gas separation actually began long before that period [5]. Gas separation processes require a membrane with high permeability and selectivity that is the asymmetric membrane [6]. Asymmetric membranes that are suitable for gas separation should have thin and dense skin layers supported by thick porous sub-layers [7–9]. Membrane structures can be classified according to the cross section microstructure as shown in Figure 1. There are six major membrane processes, which are widely used, in industrial application. They are microfiltration, ultrafiltration, reverse osmosis, electrodialysis, gas separation and pervaporation. Gas separation is known as a developing process and most gas separation membranes are of the solution-diffusion mechanism type. The key membrane performance variables are selectivity, permeability and durability. For solution-diffusion membranes, permeability is defined as the product of the solubility and diffusivity. Traditionally, there has been a tradeoff between selectivity and permeability; high selectivity membranes tend to exhibits less permeability and vice versa.

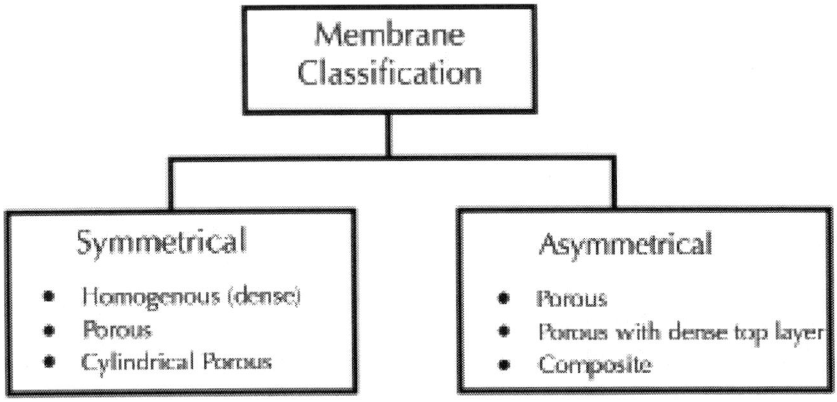

Figure 1: Membrane classifications.

MEMBRANE TECHNOLOGY FOR GAS SEPARATION

Membrane can be defined as a selective barrier between two phases, the "selective" being inherent to a membrane or a membrane process [10]. The membrane gas separation technology is over 15 years old and is proving to be one of the most significant unit operations. The technology inherits certain advantages over other methods that include compactness and light weight, low labor intensity, modular design permitting easy expansion or operation at partial capacity, low maintenance, low energy requirements, low cost, and environmental friendly.

The progress in membrane science and technology was accelerated during the 1980s by the development and refinement of synthetic polymeric membranes. Membrane gas separation emerged as a commercial process on a large scale during the 1980s. During this period, significant progress was made in virtually every aspect of membrane technology, including improvements in membrane formation process, chemical and physical structures, configurations and applications. Chemical structures coupled with subtle physical properties of the membrane material influence the permeability and selectivity of a gas. The responses of a polymeric material to permeation are strongly influenced by the polarity and steric characteristics of the polymer and permeate. The size and shape of bulky groups in both the polymer main chains and side chains determine certain fundamental properties like packing, density, and rigidity. These, in turn, influence the accessibility. An absence of such groups tends to increase the structural regularity, which favors increased density. The effect of lateral substituents on the backbone of aliphatic polyamides has a strong bearing on the gas separation properties. Bulky hydrocarbon groups force parallel chain segments further apart, thereby increasing free volume and diminishing hydrogen bonding and rigidity, and increasing the permeability [11].

In the concept of membrane science, a synthetic membrane behaves as a thin barrier between two phases through which differential transport can occur. Driving forces that facilitate this transport are pressure, concentration, and electrical potential across the R. Abedini, A. Nezhadmoghadam/Petroleum & Coal 52(2) 69-80, 2010 70 medium [12]. The transport itself is a non-equilibrium process and the separation of chemical species results from differences in transport rates through the membrane. Gas membranes are now widely used in variety of application, as shown in Table 1.

Table 1: Gas separation membrane application

Common gas separation	Application
O2/N2	Oxygen enrichment, inert gas generation
H2/Hydrocarbons	Refinery hydrogen recovery
H2/N2	Ammonia Purge gas
H2/CO	Syngas ratio adjustment
CO2/ Hydrocarbons	Acid gas treatment, landfill gas upgrading
H2O/ Hydrocarbons	Natural gas dehydration
H2S/ Hydrocarbons	Sour gas treating
He/ Hydrocarbons	Helium separation

He/N2	Helium recovery
Hydrocarbons/Air	Hydrocarbons recovery, pollution control
H2O/Air	Air dehumidification

POLYMERIC (ORGANIC) MEMBRANES

Polymeric membranes have been developed for variety of industrial applications including gas separation. For gas separation, the selectivity and permeability of the membrane material determines the efficiency of the gas separation process. Based on flux density and selectivity, a membrane can be classified broadly into two classes: (a) porous and (b) nonporous. A porous membrane is a rigid, highly voided structure with randomly distributed inter-connected pores. The separation of materials by porous membrane is mainly a function of the permeate character and membrane properties, such as the molecular size of the membrane polymer, pore-size, and pore-size distribution. A porous membrane is very similar in its structure and function to the conventional filter. In general, only those molecules that differ considerably in size can be separated effectively by microporous membranes. Porous membranes for gas separation do exhibit very high levels of flux but inherit low selectivity values. Microporous membranes are characterized by the average pore diameter d, the membrane porosity, and tortuosity of the membrane. There are several ways to prepare porous polymeric membranes, such as solution casting, sintering stretching, track etching, and phase separation. The final morphology of the membrane obtained will vary greatly, depending on the properties of the materials and the process conditions utilized. Nonporous or dense membranes have high selectivity properties but the rates of transport of gases through the medium are usually low. An important property of a nonporous dense membrane is that even permeants of similar sizes may be separated if their solubility in the membrane differs significantly. A dense membrane can be prepared by melt extrusion, where a melt is envisioned as a solution in which the polymer in both solute and solvent. In the solution-casting method, dense membranes are cast from polymer solutions prepared by dissolution of a polymer in a solvent vehicle to form a sol. This is followed by complete evaporation of the solvent after casting. Polymer membranes have gained popularity in isolating carbon dioxide from other gases [13]. These membranes are elastomers formed from crosslinked copolymers of high molecular weights. They are prepared as thin films by extrusion or casting. They demonstrate unique permeability properties for carbon dioxide together with high selectivity towards H_2, O_2, N_2, and CH_4. These elastomers are very thin membranes (5–10 m) and can be easily prepared by extrusion, solvent casting, or impregnation. The membranes are hydrophilic and the introduction of hydrophobic monomers provides regulation of water absorption. They are thermally stable up to 250°C and are biodegradable and nontoxic and their applications include food packaging, natural or industrial gas treatment of confined atmospheres (airplane cabins, air conditioned rooms), decarbonation of gases used in fuel cells, biomass treatment, medical uses.

DIFFERENT POLYMERIC (ORGANIC) MEMBRANES

The effects of flexible hetero-atoms and cyclic structures, with nonlinear orientation in the polymer backbone lead to a decreased structural regularity, on permeation behavior [14]. R. Abedini, A. Nezhadmoghadam/ Petroleum & Coal 52(2) 69-80, 2010 71 the structural regularity favors close packing of molecules which improve density and rigidity in both crystalline and amorphous phases. All these changes have some bearing on the effects of permeability and selectivity. Polyether sulfones (PES) have a relatively more regular structure and inherits a higher bulk density (1.37 g/cm^3), and an elevated glass transition Tg, (220°C). It also acquires a lower free volume when compared to polysulfone (PSF) which contains an aliphatic isopropyllidene group contributing to less regularity, lower bulk density (1.23 g/cm^3), and lower Tg (190°C). Isotropic Polyethersulfone (PESF) dense film have been proven to exhibit better selectivity for the commercially important gas pairs (CO_2/CH_4, He/CH_4, H_2/N_2, O_2/N_2) compared to bisphenol-A polysulphone and cellulose acetate [15]. Wang et al. (1990) reports that similar to PSF, PESF possesses properties of high mechanical and thermal stabilities [16]. Moreover, it is generally easy to prepare the dried asymmetric membranes by the immersion phase inversion method using water as a coagulant. This polymer has only a moderate permeability. In order to use this polymer to prepare commercially attractive gas separation membranes, fabrication of ultrathinskinned asymmetric hollow fiber membranes will be needed so as to increase permeation flux. Only a few studies on the preparation of PESF asymmetric hollow fiber membranes for gas separation have been reported in the literature. The PESF hollow fibers prepared at various spinning conditions exhibit reduced gas permeation behavior. The observed CO_2 pressure-normalized flux is less than 12GPU at 24°C. In the later 1980s, PESF hollow fiber membranes with good permeance and selectivity were prepared from spinning solution containing 1:1 molar mixtures of propionic acid: NMP and high polymer concentration (more than 35 wt%) [17]. the oxygen pressure-normalized flux was prepared to be 13.1GPU with O_2/N_2 selectivity of 5.1 at 50°C. Fritzsche et al. (1990) examined the skin layer structure of membrane [18]. Wang et al. (2000) made studies on the preparation and characterization of PESF hollow fiber membranes spun from moderate polymer concentration (25–30 wt %) and solvent systems containing various alcohols as non-solvent additives (NSA) [19]. This NSA plays a dominant role in determining membrane structure and gas separation properties. They prepared the PESF hollow fiber membranes with the best combination of gas permeability and selectivity by using ethanol as an additive in the various studies. Their studies also demonstrated that effective NSAs possess high affinities and diffusivities with the coagulant. It is also known that a certain amount of water can be added to the polymer solution before phase separation occurs. In addition, water and N, N-substituted amides (solvents) can form a complex through multiple hydrogen bonding. Therefore, the influence of water as a NSA on the structure and gas permeation properties of PESF hollow fiber membranes is of immediate interest. The authors demonstrated that siliconcoated PESF asymmetric hollow fiber membranes with high pressure-normalized fluxes and ideal selectivities for gas pairs of He and N_2 were fabricated from NMP/H_2O solvent system with mass ratio of 8.4:1. In the presence of water as an additive, the viscosity of spinning solutions increased dramatically due to the formation of NMP:H_2O hydrogenbonding complex. The observed pressure-normalized fluxes and ideal selectivities of the PESF membranes were higher than those of the hollow fiber membranes fabricated from NMP/alcohol and NMP/ propionic acid systems reported in literature.

Saha and Chakma (1995) suggested attaching a liquid membrane in a microporous polymeric membrane [20]. They immobilized mixtures of various amines such as monoethanolamine (MEA), diethanolamine (DEA), amino-methyl-propanol (AMP), and polyethylene glycol (PEG) in an icroporous polypropylene film and placed it in a permeator. They tested the mechanism for the separation of carbon dioxide from some hydrocarbon gases and obtained separation factors as high as 145. Fuertes (2001) reported that when the size of micropores in the carbon membranes is in the 3–5 Ǻ range, gas molecules with effective diameters smaller than 4Ǻ display significant differences in gas diffusivity, and mixtures of these gases can be effectively separated according to a molecular sieving mechanism [21]. In this case, the gas transport rate through the membrane depends on the effective size of the gas molecules instead of the adsorption effects. Membranes with these characteristics are identified as Molecular Sieve Carbon Membrane (MSCM). Polymers (polyvinylidene chloride, polyimides, polyfurfuryl alcohol, phenolic regins) that are used to prepare MSCM are initially cross-linked or become cross-linked during pyrolysis. This prevents the formation of large graphite-like crystals during carbonization and leads to the formation of disordered structures (non-graphiting carbons) with a very narrow porosity formed by R. Abedini, A. Nezhadmoghadam/ Petroleum & Coal 52(2) 69-80, 2010 72 micropores in the 3–5 Ǻ range (molecular sieve carbons). In this sense, an MSCM can be described as a thin molecular sieve carbon film. If the micropores in the carbon membrane are enlarged from 3 to 5 Ǻ, gas permeation through the membranes with enlarged micropores is governed by the adsorption strength instead of molecular size. This kind of membrane would be effective for separation of non-adsorbable or weakly adsorbable gases from adsorbable gases.

INORGANIC MEMBRANES

The efficiency of polymeric membranes decreases with time due to fouling, compaction, chemical degradation, and thermal instability. Because of this limited thermal stability and susceptibility to abrasion and chemical attack, polymeric membranes have found application in separation processes where hot reactive gases are encountered. This has resulted in a shift of interest toward inorganic membranes. Inorganic membranes are increasingly being explored to separate gas mixtures. Besides having appreciable thermal and chemical stability, inorganic membranes have much higher gas fluxes as compared to polymeric membranes. There are basically two types of inorganic membranes - dense (nonporous) and porous membranes. Examples of commercial porous inorganic membranes are ceramic membranes, such as alumina, silica, titanium, and glass and porous metals, such as stainless steel and silver. These membranes are characterized by high permeabilities and low selectivities. Dense inorganic membranes are very specific in their separation behaviors, for example Pdmetal based membranes are hydrogen specific and metal oxide membranes are oxygen specific. Palladium and its alloys have been studied extensively as potential membrane materials. Air products and Chemical Inc. developed the selective surface flow (SSF) membranes. It consists of a thin layer (2–3 mm) of nanoporous carbon supported on a macroporous alumina tube [22]. The effective pore diameter of the carbon matrix is 5–7 Ǻ [23]. The membrane separates the components of a gas mixture by selective adsorption– surface diffusion–desorption mechanism [24]. The larger polar molecules of a feedgas mixture at the high-pressure zone of the membrane are selectively adsorbed on the carbon pore walls followed by their selective diffusion along the pore walls to the low pressure side of the membrane where they desorb into the low pressure gas phase. Thus, a gas stream enriched in the smaller and less polar components of the feed-gas stream mixture is produced as a low pressure effluent gas from the Selective Surface Flow (SSF) membrane system. However, a gas stream enriched in the larger and more polar components of the feed-gas mixture is produced as a low-pressure effluent gas. The performance of the SSF membrane for the separation of hydrocarbons from H_2, CO_2 from H_2, CO_2 and CH_4 from H_2, H_2S from H_2, and H_2S from CH_4 are found in the literature [25]. Gas separation by means of microporous carbon membrane is based on interaction differences existing between components of a gas mixture with respect to the carbon membrane. Materials that are thermally stable at temperatures above 400°C include ceramic membranes that are used to separate hydrogen from gasified coal [26]. With ceramic materials, very high separation factors have been achieved based on the ratios of individual gas permeances. This work focuses on the development of alumina membranes having nanopores that can separate gases based on the molecular size. A layer of alumina having a mean pore diameter of about 7nm is deposited on a stainless steel tube. Methods have been developed to reduce the pore diameter to values in the range between 0.5 and 3.0 nm. The membranes having the smallest estimated mean pore diameters have shown very high separation factors for helium and hydrogen with other gases. These separation factors are determined from pure gas separations in these pore diameter ranges. For pure molecular flow, separation factors should increase with decreasing pore size, but in many cases, the separation factor decreases. This decrease occurs because of adsorption and surface flow of the heavier molecule. The separation factor continues to decrease until the pore size becomes small enough that the entrance into the membrane is restricted for the larger molecule. At this point, the separation factor increases rapidly as pore size is decreased. It is observed that the separation factors are directly proportional to temperature. This is because with increase of temperature, an unexpected large increase of permeance of hydrogen and helium is observed.

MIXED MATRIX MEMBRANES (MMM_S)

An upper limit for the performance of polymeric membranes in gas separation was predicted by Robeson [27] in early 1990. The performance of various membrane materials available for the separation of O_2/N_2 is depicted in Figure 2. The figure presents the permeability of the fast gas O_2 on the abscissa on a logarithmic scale and the O_2/N_2 selectivity on the ordinate, again on a logarithmic scale. For the polymeric materials, a rather general trade-off exists between permeability and selectivity, with an "upperbound" evident in Figure 2.

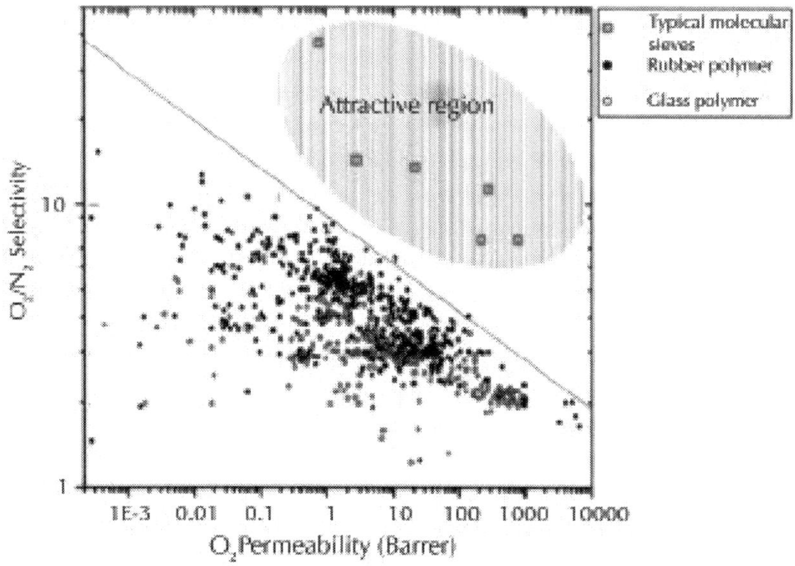

Figure 2: Relationship between the O_2/N_2 selectivity and O_2 permeability for polymeric membranes and inorganic membranes [27] (the dots indicates the performance of polymeric materials).

When materials with separation properties near this limit were modified based on the traditional structure – property relation, the resultant polymers have permeability and selectivity tracking along this line instead of exceeding it. On the other hand, as may be seen in Figure 2, the inorganic materials have properties lying far beyond the upper-bound limit for the organic polymers [28–32]. Though tremendous improvements had been achieved in tailoring polymer structure to enhance separation properties during the last two decades, further progress exceeding the trade-off line seems to present a severe challenge in the near future. Similarly, the immediate application of inorganic membranes is still seriously hindered by the lack of technology to form continuous and defect-free membranes, the extremely high cost for the membrane production, and handling issues (e.g., inherent brittleness) [33,34]. In view of this situation, a new approach is needed to provide an alternate and cost-effective membrane with separation properties well above the upper-bound limit between permeability and selectivity. The latest membrane morphology emerging with the potential for future applications involves MMM, consisting of organic polymer and inorganic particle phases, as shown schematically in Figure 3. The bulk phase (phase A) is typically a polymer; the dispersed phase (phase B) represents the inorganic particles, which may be zeolite, carbon molecular sieves, or nano-size particles. MMMs have the potential to achieve higher selectivity, permeability, or both relative to the existing polymeric membranes, resulting from the addition of the inorganic particles with their inherent superior separation characteristics. At the same time, the fragility inherent in the inorganic membranes may be avoided by using a flexible polymer as the continuous matrix.

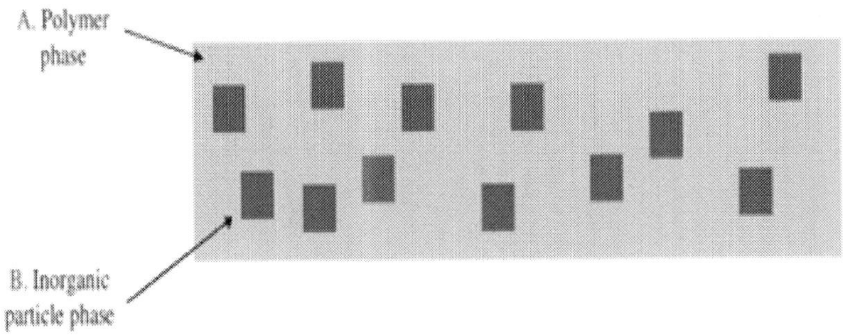

Figure 3: Schematic of a mixed matrix membrane (MMM).

The investigation of MMMs for gas separation was first reported in 1970s with the discovery of a delayed diffusion time lag effect for CO_2 and CH4 when adding 5A zeolite into rubbery polymer polydimethyl siloxane (PDMS) [35]. In this work, Paul and Kemp found that the addition of 5A into the polymer matrix caused very large increases in the diffusion time lag but had only minor effects on the steady-state permeation. Researchers at UOP were the first to report that that mixed matrix systems of polymer/adsorbent might yield superior separation performance to that of pure polymeric system [36]. They observed an enhanced O_2/N_2 selectivity from 3.0 to 4.3 when increasing silicalite content in the polymer cellulose acetate (CA) matrix. The concept of MMM has been also demonstrate at UOP LLC in the mid-1980s using CA/silicalite MMMs for CO_2/H_2 separation [37]. In the demonstration, a feed mixture of 50/50 (mol%) CO_2/H_2 with a differential pressure of 50 psi was used. The calculated separation factor for CO_2/H_2 was found to be 5.15±2.2. In contrast, a CO_2/H_2 separation factor of 0.77±0.06 was found for CA membrane. This indicates that silicalite in the membrane phase has reversed the selectivity from H_2 to CO_2.

Many studies have demonstrated that the remarkable separation properties of MMMs accord with this design, exhibiting performance well beyond the intrinsic properties of the polymer matrix. The most prominent work involved the application 4A zeolite [38-40]. Figure 4 shows the O_2/N_2 transport properties of these membranes in comparison with the Robeson, 1991 O_2/N_2 upperbound limit curve, indicating that MMMs are promising candidates for the next generation of membranes.

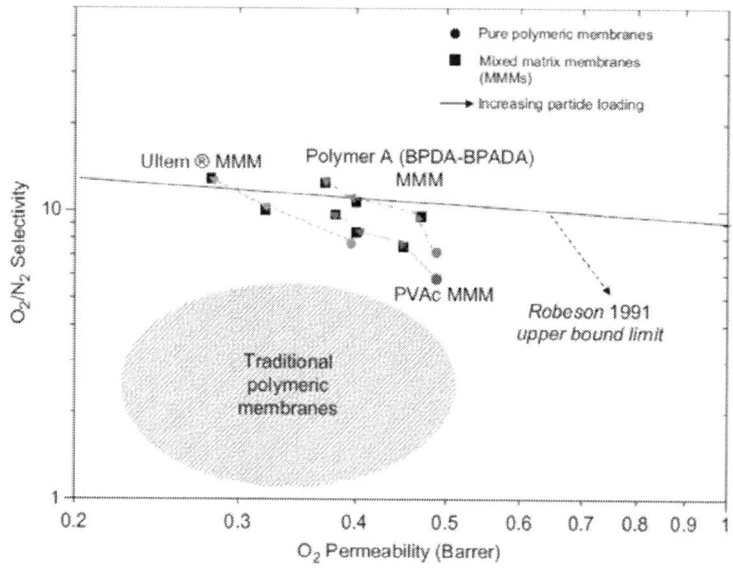

Figure 4: Mixed matrix membranes performance compared to the Robeson, 1991 upperbound limit [54, 56].

MECHANISMS FOR GAS SEPARATION

Various mechanisms for gas transport across membranes have been proposed depending on the properties of both the permeant and the membrane. These include Knudsen diffusion, the molecular sieve effect and a solution diffusion mechanism. Most of these models, however, have been found to be applicable only to a 1imited number of gas/ material systems. Different mechanisms (a–d) may be involved in the transport of gases across a porous membrane gives a schematic representation of the mechanisms for the permeation of gases through porous as well as dense membranes (Baker, 1995).

Mechanism (a)

The permeation of gas through porous membrane consists of Knudsen diffusion and Poiseulle flow. The properties of Knudsen to Poiseuille flow are governed by the ratios of the pore radius (r) to the mean free path (λ) of the gas molecules. The mean free path is given by

$$\lambda = \frac{3\eta\left(\pi RT\right)^{0.5}}{2P\left(2M\right)} \tag{1}$$

Where η is the viscosity of the gas, R the universal gas constant, T the temperature, M the molecular weight, and P the pressure. If $\lambda / r <<$ l viscous or Poiseuille flow predominates then the gas flux through the pore is described by the following expression:

$$G_{vis} = \frac{r_2\left(P_1 - P_2\right)}{16LuRT} \tag{2}$$

where, r is the pore radius, P_1 the partial pressure of the gas on the feed side, P_2 the partial pressure of the permeate side, L the pore length, u the gas viscosity, Gives the viscous flow, and R the universal gas constant. In Knudsen flow, λ / r is considerably less than unity, and therefore, there are more collisions with the pore walls than with other gas molecules. At every collision with the pore walls, the gas molecules are momentarily absorbed and then reflected in a random direction. As there is less number of collisions among molecules than pore walls, each molecule will move independent of others. Hence, the separation is achieved because different gaseous species are moving at different velocities.

The gas flux in such a case is described by the expression:

$$G_{mol} = \frac{8r\left(P_1 - P_2\right)}{3L\left(2\pi MRT\right)^{0.5}} \tag{3}$$

where Gmol is the molecular flow of the gas, r the pore radius, P_1 the partial pressure of the gas on the feedside, P_2 the partial pressure of the gas on the permeate side, L the pore length, M the molecular weight, R the gas constant, and T absolute temperature. For Knudsen flow, the selectivity ratio or the separation factor for binary gas mixtures can be estimated from the square root of the ratio of the molecular weights. Knudsen separation can be achieved with membranes having pore sizes smaller than 50 nm. Based on Knudsen flow, separation factors for several gas pairs have been calculated and they represent ideal separation factors.

Mechanism (b)

In order to function as a molecular sieve, membranes must have pore diameters that are in between those of the gas molecules to be separated. Separation factors greater than 10 should be achievable as the pores become smaller than approximately 0.5 nm. If the membrane has pore sizes between the diameter of the smaller and the larger gas molecules, then only the smaller molecule can permeate and a very high separation would be achieved. In practical situations, there will be a distribution of pore size in the membranes, and thus the gas permeability is actually influenced by a combination of transport mechanisms. From a practical standpoint, as the pore size decreases, the membrane porosity is expected to decrease, resulting in a lower gas flow through the membrane. Therefore, the pore size and porosity must be balanced to produce an efficient membrane.

Mechanism (c)

Gas separation can also be affected by partial condensation for some component of a gas mixture in the pores, with the exclusion of others, and the subsequent transport of the condensed molecules across the pore.

Mechanism (d)

Selective adsorption of the more strongly adsorbed components of a gas mixture onto the pore surface followed by the surface diffusion of the adsorbed molecule across the pore can also facilitate the separation of gases. A proper concentration gradient for the diffusing species must be imposed across the porous membrane to provide the driving force for transport by all four mechanisms.

The selectivity of separation achieved by Knudsen mechanism (a) is generally very low. Mechanism (b) exhibits high selectivity and high permeability for the smaller components of a gas mixture. Mechanism (c) requires the pore size of the membrane to be in the mesoporous size range (diameter>30) so that condensation of the component of a gas mixture can take place. A very high selectivity of separation of the condensable component can be achieved by this mechanism. The condensation partial pressure of the component at the system pressure, the pore size and the geometry of the membrane limit the extent of removal of that component from the gas mixture. Mechanism (d) provides the most flexible and attractive choice for the practical separation of gaseous mixtures. The reason is that the separation selectivity is determined by the preferential adsorption of certain components of the gas mixture on the surface of the membrane pores as well as by the selective diffusion of the absorbed molecules.

In meso- and microporous media, as in the case of inorganic membranes, if the relative pressure is increased, the adsorbed and capillary condensed materials permeate together. The porous media behave, as a semipermeable membrane through which the adsorbed permeate will flow freely while the weakly adsorbed component will be blocked. Thus, both the pore size and the physical–chemical nature of the pore surface play key roles in determining the separation efficiency of these membranes. Consequently, appropriate molecular engineering of their surface chemistries can alter the properties of the membranes.

The transport of gases through a dense polymeric membrane is usually described by the solution- diffusion mechanism. The important feature of dense membranes that are used in separation applications is the ability to control the permeation of different species. In the solution diffusion mechanism, the permeants dissolve in the membrane material and then diffuse through the membrane down a concentration gradient.

A separation is achieved between the different permeants because of differences in the amount of material diffusing through the membranes.

The solution-diffusion mechanism is considered to consist of three steps: (1) the absorption or adsorption at the upstream boundary, (2) activated diffusion (solubility) through the membrane, and (3) desorption or evaporation on the other side. These solution-diffusion mechanisms are driven by a difference in thermodynamic activity at the upstream and downstream faces of the membrane as well as the interacting forces between the molecules of the membrane material and permeate molecules. The activity difference causes a concentration difference that leads to diffusion in the direction of decreasing activity. As in microporous membranes, differences in permeability of dense membranes also result from differences in the physical–chemical interactions of the gas species within the polymer. The solution-diffusion model assumes that the pressure within a membrane is uniform and the chemical gradient across the membrane is expressed only as a concentration gradient.

The schematic diagram shown in Fig. 5 illustrates mass transfer across a regular membrane and a liquid membrane. In a regular membrane, the solute a passes through the membrane under pure diffusion mechanism, leaving the other gas behind. In the liquid membrane, the enhanced effect of having the liquid solvent, B, can be seen by the additional flux of species AB. This system, with the additional flux, is known as facilitated transport. Without the effects of the liquid, mass transfer would be reduced to that of the regular diffusion membrane system whereby the total flux only consists of the flux of species a including pressure, concentration, and electrical potential across the membranes.

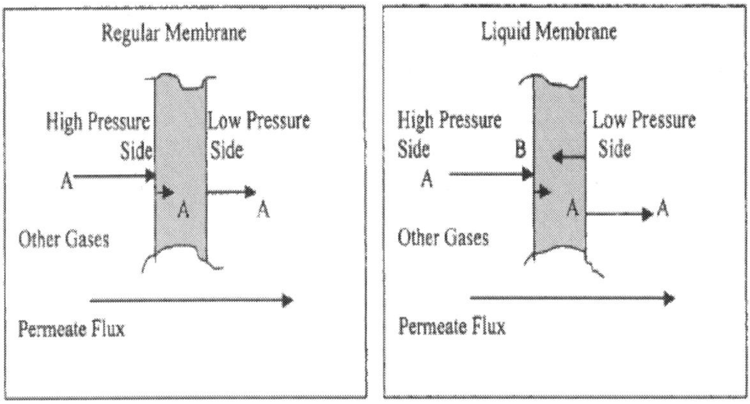

Figure 5: Faciliated transport in a membrane.

Pressure difference across the membrane can facilitate RO (reverse osmosis), ultrafiltration, micro-filtration, gas separation, and pervaporation. Temperature difference across the membrane can facilitate distillation, whereas concentration difference can be used for dialysis and extraction. Electro-dialysis can also be performed with the help of membranes when an electric potential across the membrane is maintained. Differential transport occurs when the membrane restricts the transport of different species in some specific way. The transport is a non-equilibrium process and the separation of chemical species results from differences in transport rates through the membranes. Membrane separation is considered a non-equilibrium process because it is based on the selective permeation rate of the feed components. If a membrane system is allowed to move toward equilibrium, permeation would continue until the pressure and concentration of gases on both sides are equal (Koros et al., 1996) [12].

The two main parameters defining the performance of a membrane are separation factor and permeance. In a binary system consisting of gases "a" and "b" with gas "a" as the fast permeating gas, separation factor is defined as the concentration ratio of "a" to "b"in the permeate, and X_a and X_b are the concentration of the gases "a" and "b" in the feed.

$$\alpha_{a/b} = \frac{(Y_a / Y_b)}{(X_a / X_b)} \tag{4}$$

Where α is the separation factor, Y_a and Y_b are the concentration of the gases "a" and "b" in the permeate and X_a and X_b are the concentrations of gases "a" and "b" in the feed. Permeance is defined as the volume of the feed passing through a unit area of membrane at unit time and under unit pressure gradient and is expressed as:

$$P = \frac{V}{At\Delta p} \tag{5}$$

Where P is permeance, V is volume of gas, A is area, t is time, and Δ_p is pressure difference (Tabe-Mahammadi, 1999) [41].

CONCLUSIONS

Membrane systems are highly applicable for separation of gases from a mixture. Continuing enhancements in technology developments of membrane systems is a natural choice for the future. Gas separation modules operate on the basis of selective permeation. The technology takes advantage of the fact that gases dissolve and diffuse into polymeric materials. If a pressure differential is set up on opposing sides of a film (membrane), transport across the film (permeation) will occur. The product of a solubility coefficient and a diffusion coefficient determines the rate of permeation. High performance zeolite materials as a filler for mixed-matrix membranes can further take gas separation membranes into uncharted areas. Membrane separations employing mixed-matrix materials have the potential to meet demanding gas separation in a world with shrinking energy resources and increasing environmental concerns. Any membrane will separate gases to some extent. However, proper selection of the polymetric material comprising the membrane is extremely important as it determines the ultimate performance of the gas separation module. Membrane productivity for a given base material is determined by three factors: partial pressure difference across the membrane, membrane thickness, and membrane surface area. The partial pressure difference is typically determined by the specifics of the application. The membrane manufacturer can address the other two factors determining membrane productivity.

REFERENCES

1. Tarmoon, I. O. (1999). Middle East Oil Show. Feb. 20–23.
2. Sirkar, K.K. and Winston Ho, W.S., Part I and Part II in Membrane Handbook. New York: Van Nostrand Reinhold, (1992), 1-25.
3. Mark, H.F., Bikare, W.M., Overberger, C.G., and Menges, G. 1985. Encyclopedia of Science and Engineering, 9 (1985), 50-58
4. Pandey, P. and Chauhan, R.S. 2001. Membrane for gas separation. Progress in Polymer Science, 26, (2001) 853-893.
5. Baker, R.W., Membrane Technology and Application. Membrane Technology and Research, Inc.McGrow-Hill, New York (2000).
6. Lin, F.C., Wang, D.M. and Lai, J.Y., Asymmetric TPX membranes with high gas flux, Journal of Membrane Science, 110 (1996) 25-36.
7. Wang, D., Li, K. and Teo, W.K., Polyethersulfone hollow fibre gas separation membranes prepared from nmp/alcohol solvent systems. Journal of Membrane Science, 115, (1996), 85-108.
8. Pinnau, I., Wind, J. and Peinemann, K.V., Ultrathin multicomponent polyethersulpone membranes for gas separation made by dry/wet phase inversion. Ind. Eng. Chem. Res., 29, (1990) 2028- 2032.
9. Peinemann, K.V. and Lauenburg, Method for producing an integral asymmetric gas separation membrane and the resultant membrane. US Patent. 4, 746, 333 (1988).
10. Mulder, M., Basic Principles of Membrane Technology. Netherlands: Kluwer Academic Publishers (1996).
11. Pandey, P., Chauhan, R. S., Membranes for gas separation, Progress in Polymer Science, 26 (6), (2001), 853-893.

12. Koros, W. J., Ma, Y. H., Shimidzy, T. J., The passing of a giant, Journal of Membrane Science, 120, (2), (1996), Page i.
13. Gramain, P., Sanchez, J., Membrane Pour La Separation Selective Gazeuse. French Patent. WO 022245, (March, 2002).
14. Zimmerman, J., Polyamide in Encyclopedia of Polymer Science and Technology. New York: John Wiley and Sons (1984).
15. Ellig, D. L., Althouse, F. P., McCandless, C., Concentration of methane from mixtures with carbon dioxide by permeation through polymeric films, Journal of Membrane Science, 6, (1980), 259-263.
16. Wang, D., Li, K., Teo, W. K., Preparation and characterization of high-flux polysulfone hollow fibre gas separation membranes, Journal of Membrane Science, 204, (2002), 247-256
17. Kesting, R. E., Fritzsche, A. K., Murphy, M. K., Handermann, C. A., Cruse, C. A., Malon, R. F., Process for Forming Asymmetric Gas Separation Membranes having Graded Density Skins. US Patent, 871 (April, 1989).
18. Fritzsche, A. K., Cruse, C. A., Murphy, M. K., Kesting, R. E., Polyethersulfone and polyphenylsulfone hollow fiber trilayer membranes spun from Lewis acid: base complexes Structure determination by SEM, DSC, and oxygen plasma ablation,Journal of Membrane Science, 54, (1990), 29-50.
19. Wang, D., Li, K., Teo, W. K., Highly permeable polyethersulfone hollow fiber gas separation membranes prepared using water as non-solvent additive, Journal of Membrane Science, 176, (2000), 147-158.
20. Saha, S. and A. Chakma, Selective CO2 Separation from CO2/ C2H6 mixture by mobilized DEA/ PEG membranes, Journal of Membrane Science, 98, (1995), 157-171.
21. Fuertes, A. B. Effect of air oxidation on gas separation properties of adsorptionselective carbon membranes, Carbon, (39), (2001), 697-706.
22. Rao, M. B., Sircar, S., Golden, T. C., Gas Separation by Adsorbent Membranes. US Patent 5, 104, (1992), 425.
23. Rao, M. B., Sircar, S., Performance and pore characterization of nanoporous carbon membranes for gas separation, Journal of Membrane Science, 110, (1996), 109-118.

24. Rao, M. B., Sircar, S., Nanoporous carbon membranes for separation of gas mixtures by selective surface flow, Journal of Membrane Science, 85, (1993), 253-264.
25. Parillo, D. J., Thaeron, C., Sircar, S., Separation of bulk hydrogen sulfide-hydrogen mixtures by selective surface flow membrane, AIChE Journal, 43, (1997), 2239-2245.
26. Fain, D. E., Marshall, B. B., Raj, T., Addock, K. D., Powell, L. E., In: 15th Annual Conference on Fossil Energy Materials, Oak Ridge, Tennessee, April 30–May 2 (2001).
27. Robeson LM. Correlation of separation factor versus permeability for polymeric membranes, Journal of Membrane Science, 62 (1991), 165–85.
28. Singh AS, Koros WJ. Air separation properties of flat sheet homogeneous pyrolytic carbon membranes, Journal of Membrane Science, 174, (2000), 177–88.
29. Feuters AB, Centeno TA. Preparation of supported carbon molecular membranes, Carbon, 37, (1999), 679–84.
30. Kim YK, Lee JM, Park HB, Lee YM. The gas separation properties of carbon molecular sieve membranes derived from polyimides having carboxylic acid groups, Journal of Membrane Science, 235, (2004), 139–46.
31. Park HB, Kim YK, Lee JM, Lee SY, Lee YM. Relationship between chemical structure of aromatic polyimides and gas permeation properties of their carbon molecular sieve membranes, Journal of Membrane Science, 229, (2004), 117–27.
32. Tin PS, Chung TS, Kawi S, Guiver MD. Novel approaches to fabricate carbon molecular sieve membranes based on chemical modified and solvent treated polyimides, Micropor Mesopor Mater, 73, 2004, 151–60.
33. Saracco G, Neomagus HWJP, Versteeg GF, Van Swaaij WPM. High-temperature membrane reactor: potential and problems, Chem Eng Sci, 54, (1999), 1997–2017.
34. Caro J, Noack M, Kolsch P, Schaefer R. Zeolite membrane-state of their development and perspective, Micropor Mesopor Mater,38, (2000), 3–24.
35. Paul DR, Kemp DR. The diffusion time lag in polymer membranes containing adsorptive fillers, J Polym Sci: Polym Phys, 41, (1973), 79–93.

36. Kulprathipanja S, Neuzil RW, Li NN. Separation of fluids by means of mixed matrix membranes, US patent 4740219, (1988).
37. Kulprathipanja S, Neuzil RW, Li NN. Separation of gases by means of mixed matrix membranes, US patent 5127925, (1992).
38. Mahajan R, Koros WJ. Factors controlling successful formation of mixed-matrix gas separation materials, Ind Eng Chem Res, 39, (2000), 2692–2696.
39. Mahajan R, Koros WJ. Mixed matrix membrane materials with glassy polymers. Part 1, Polym Eng Sci, 42, 2002, 1420–1431.
40. Baker, R. W., Membrane technology. Kirk Othmer Encyclopedia of ChemicalTechnology, Singapore: Wiley, 16 (1995).
41. Abdulreza Tabe-Mohammadi, A Review of the Applications of Membrane Separation Technology in Natural Gas Treatment Separation Science and Technology, 34, (1999), 2095-2111.

Chapter 8

A Comparative Study of Liquefied Natural Gas: An Overview

Muhammad Khan Memon, Saleem Qadir Tunio, Khalil Rehman Memon, and Arshad Ahmed Lashari

Petroleum Engineering Department, Universiti Technologi PETRONAS, Bandar Seri Iskandar 31750, Tronoh Perak Darul Ridzwan, Malaysia

ABSTRACT

Natural gas is the world's fastest growing fuel and being produced by many countries of the world in the commercial quantities. Increasing natural gas price and new development in the technologies, liquefied natural gas industry is economically attractive in the major gas exporting countries. Liquefied Natural Gas (LNG) is an important energy source and continued to contribute the growth of natural gas industry. The new advance LNG technology is used for natural gas transportation for long distances. LNG can be transported by the large insulated cryogenic tankers at affordable cost. This study presents an overview of LNG liquidation facilities from natural gas as feed gas to LNG storage and transport. The main objective of the study is to highlight the current data for reviewers on LNG world market, mainly on LNG production, supply, demand, price and new development of LNG plants. The technology is growing gradually with increasing number of LNG consuming countries in overall the world. In the near future, LNG price may be affected by the advanced shale gas production in the United States of America and China. Australia becomes the world second largest exporter of LNG market after Qatar. Australia will increase LNG supply by 15 Bcf/day from 2014 and accounting for 25% of world LNG production by 2030. Global LNG production forecast will be reached 540 Bcm by 2020 and LNG trade will be reached 425 Mtpa by 2025. New countries are interested to enter in the LNG world market as importers and exporters.

INTRODUCTION

LNG Fundamentals

The first LNG was imported in limited quantity by Japan in 1960. The LNG technology was developed in Japan after oil shocks of 1970 (Small, 2005). With increasing demand for the clean burning fuel, LNG projects are increasing globally. Nigeria, Norway, Egypt, Trinidad and Qatar imported the LNG from their new plant worldwide. In the year 2000-2010 it was approximately doubled 175 Mtpa. A single train LNG plant may cost 1.5 billion USD and consume 6 to 8% of the inlet gas as fuel. Impurities such as Carbon dioxide (CO_2), Hydrogen Sulfide (H_2S), Water vapor and heavier hydrocarbon compounds in natural gas must be removed. Therefore the natural gas (feed gas) must go through a series of pretreatment steps. This can prevent the damage to liquefaction equipment and meet the specification of customer (Choi, 2010).

LNG Processes

When the distance between source and customer is too long (for example 1500 km by sea or 5000 km over land), it is uneconomical to transport the Natural Gas in its original form. Therefore, the Natural Gas (feed gas) is then compressed as Liquefied Natural Gas by injecting at high pressure. Feed gas must be free from condensates, impurities and heavier hydrocarbons. The feed gas is further cooled under refrigeration process at temperature between -159 to -162°C. At this temperature the gas condenses to a liquid state completely. Flashed vapors and boil off gas are recycled within the process (Mokhatab and Economides, 2006). The pressurized LNG is further sub cooled in one or more stages to facilitate storage at slightly above atmospheric pressure. Compared to gaseous state, LNG which is in liquid form only takes up about 1/600 spaces needed by the natural gas (MEO Australia, 2013). LNG is converted to its gaseous state (natural gas) through a re-gasification facility at a receiving terminal where this energy source can be supplied through the pipeline. Figure 1 presents the flow diagram of LNG processes by Barclay and Shukri (2005).

LNG Storage and Transportation

The capacity of LNG storage is based on the volume of tanker plus four day production. Storage is the expensive item for LNG plant. Majority of LNG tankers having capacity of approximately 140000 m^3 were built during the year 2000-2010. They carry the gas 3.1 Bscf (87 million m^3) or 512000 BOE. Super LNG tankers having capacity of 265000 m3 were designed for long trade from Qatar to United States of America. In order to meet the requirement of tanker load, most of the LNG plants having two or more storage tanks (Choi, 2010). The LNG physical properties allow for its long distance transport from one country to another by ship across ocean to markets and its local distributions. Reducing the LNG transportation costs using cargo ships is an advantage over the construction of long distance pipeline. LNG storage in the tankers is the solution of underground gas storage where geological subsurface conditions are not suitable. LNG storage in the tankers can reduce the cost of gas supply through long distance pipeline (Mokhatab and Economides, 2006).

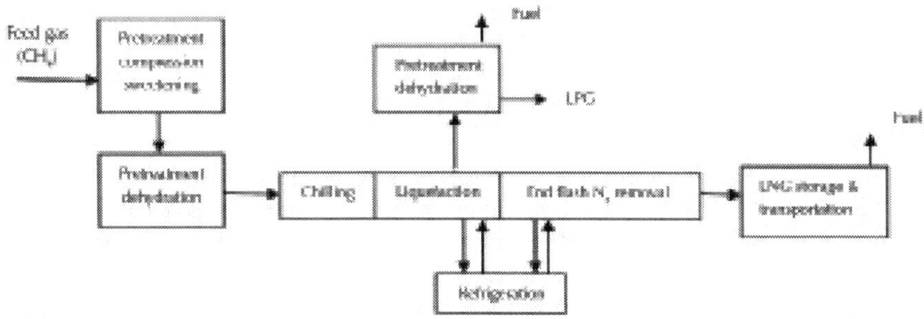

Figure 1: LNG processes (Barclay and Shukri, 2005).

Table 1: LNG and gas export by USA in 2012 (US Energy Information Administration, 2013)

LNG export (Mcf)		**Natural gas export (Mcf)**	
Brazil	8154	Canada	970846
India	3004	Mexico	619802
Japan	14379	Total	1590648
Portugal	2618		
Canada	2		
Mexico	153		
Total	28298		

LITERATURE REVIEW

Worldwide LNG Industry

In this study, some countries are selected for comparative study on LNG industry. The main objective of the study is to highlight the current data for reviewers on LNG world market, mainly on LNG production, supply, demand, price and new development of LNG plants.

United States of America (USA)

USA exported the LNG and natural gas to the Mexico and other countries of the world. Table 1 presents the LNG and Gas exported by the USA to the countries in 2012. In March 2013, USA exported the LNG to the Mexico (19 Mcf) and Canada (4 Mcf). Also, Natural Gas was supplied through the pipeline to the Mexico (56107 Mcf) and Canada (92889 Mcf) in the same month (US Energy information Administration, 2013). The first LNG shipment from USA to Asia is expected by the end of 2015, after completing the first four trains of Sabine Pass. USA will export the total LNG capacity of 18 Mtpa by 2018. Sabine Pass USA, LNG with capacity of 26.8 Bcm will be expected to complete in 2015. Freeport (14.5 Bcm) and Lake Charles (20.7 Bcm) will be completed in 2015 (Qatar LNG to Asia, 2012).

Qatar

Qatar is the world's foremost LNG exporter. Nearly 3600 Bcf of LNG was exported to the United Kingdom, Japan, South Korea and India in 2011. In 2012, about 15.7 MT LNG was exported to Japan and 10.8 MT to South Korea. About 63% LNG from Qatar was exported to Asian countries including Japan. On demand side seven importing countries out of 26 (China, Japan, South Korea, India, Taiwan, Spain and United Kingdom) attracted 81% of total LNG (US EIA Qatar, 2013; GIIGNLG, 2012).

According to the published data in January 2013 by Oil and Gas Journal, Qatar holds 13% of total world gas reserves (890 Tcf). This is the third largest total gas reserves in the world behind Russia and Iran. The major gas reserves are located in the offshore North field. This field is the key to Qatar's natural gas development and production plans (US EIA Qatar, 2013).

Qatar will maintain LNG export for next few years. New exporter competitors such as Australia, North America, Africa and Mediterranean push ahead with new LNG market and they will target Asian customers with new LNG prices (Qatar LNG to Asia, 2012).

Australia

Australia is a leader in CBM-to-LNG development. Estimated gas reserves are 3.8 Tcm including 400 Bcm of Coal Bed Methane (CBM) or Coal Seam Gas (CGS). CBM production is driven mostly by LNG export projects. Australia is the most growing LNG producer in the world. The LNG production will reach 75.8 Bcm/year by 2016. This is the world's second largest LNG exporter after Qatar. Australia will be in the position to export LNG with capacity between 60-100 Mtpa by 2020 (Qatar LNG to Asia, 2012). LNG production capacity from North West shelf Darwin is 19.6 Mtpa (26.7 Bcm). The production capacity of Pluto LNG train is 6.5 Bcm. Gorgon LNG production capacity of 6.8 Bcm will be expected to complete by 2014. Prelude LNG terminal is the world's first floating LNG terminal and will be started with capacity of 4.9 Bcm in 2017. Gladstone LNG project with capacity of 10.6 Bcm will be expected to compete in 2015 (International Energy Agency, 2011).

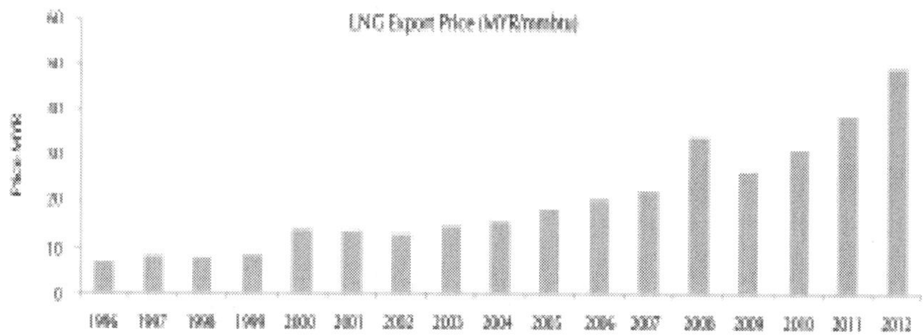

Figure. 2: LNG export price in Malaysia (Kuncinas, 2013; Koh, 2013).

Russia

Russia is the largest gas supplier through pipeline (The Voice of Russia, 2012). A new LNG plant study was carried out by Japan and Russia. Vladivostok with producing capacity of 6.8 Bcm will be constructed from 2017. Sakhalin 1, 2 (Train II) and 3 with capacities (7.5, 6.5 and 7.5 Bcm, respectively) will be expected to start in 2015. Yamal LNG project with capacity of 20.4 Bcm will be started in 2017. Shtokman

LNG project with capacity greater than 10.2 Bcm will be started in 2018 (International Energy Agency, 2011).

Singapore

Singapore has no domestic gas production. Natural gas was imported from Malaysia through pipeline since 1992. In 2010 the gas consumption was 8.7 Bcm. Singapore LNG Corporation (SLNG) built and operated its own LNG terminal with storage capacity of 4.1 Bcm. In the year 2010, the Singapore Energy Market Authority has announced the third storage tank with capacity of 4.1 Bcm at the SLNG terminal. The storage capacity of this terminal will be doubled (8.2 Bcm) in 2014 (OECD/IEA, 2013).

Malaysia

LNG in Bintulu was the first major gas project in Malaysia started in 1983. This was the first world scale LNG project. The project was a joint venture between PETRONAS, Shell Gas BV and Mitsubishi Corporation of Japan (Abdullah, 1998). Major importers of Malaysia's LNG are Japan, Korea and Taiwan. In 2009, Malaysia was the world's second largest exporter of LNG. However, declining the gas recovery from existing fields and rising demand from country's power plant, industries, positioned Malaysia one step back in the world. Malaysia exported over (1 Tcf) of LNG in 2010 (10% of the world LNG export). In 2010 Malaysia was the third largest LNG export country in the world (OECD/IEA, 2013). Malaysia's overall LNG production capacity stands at 25.7 Mtpa. Malaysia aims to be Asia's leading hub in LNG production and business. Figure 2 presents the Malaysia's LNG export price. The LNG price was increased from 1996 to 2008. The price was shown decline in 2009, due to the infrastructure development and re-gasification projects initiated by the PETRONAS (Kuncinas, 2013; Koh, 2013). Malaysia is on one step closer to become the world's largest LNG producer.

The largest LNG complex in the world was facilitated at Bintulu Sarawak Malaysia. Total liquefaction capacity of this LNG complex was 1.7 Tcf/year from eight production trains. The ninth LNG train will be added to Bintulu Sarawak Complex on Borneo, announced by

PETRONAS in February 2012 (OECD/IEA, 2013). The capacity of ninth train will be 3.6 million tones/year and bring the total capacity at the site to 27.6 million tones/year. The plant will start at the end of 2015. The feed gas for the new train will be produced by offshore fields of Sarawak (Oil and Gas Journal, 2013).

The first LNG re-gasification terminal to receive and store LNG with two floating storage units (Tenaga Empat and Tenaga Satu) is located at sungai udang port in Melaka Malaysia. The terminal linked with Jetty regasification unit and will start this year. The maximum capacity of this terminal is 3.8 mmtpa. For import and re-gasification of LNG a new LNG terminal project by PETRONAS is under construction at Pengerang, Johor. The terminal capacity is 3.8 mmtpa. This terminal will be expected to complete in 2016 (The Edge Malaysia, 2013; Petronas Malaysia MLNG, 2013; GIIGNLG, 2012).

A new Asia's largest LNG trading hub will be established with US$1.3 billion at Pengerang Integrated Petroleum Complex in Malaysia. This terminal will allow multiple LNG users to store and trade the product and will be the first independent LNG terminal in the Asia by 2020 (PETRONAS Malaysia MLNG, 2013).

PETRONAS Malaysia signed some of the LNG import agreements to meet the increasing domestic gas market demand. First agreement was signed in May 2011 with France's GDF Suez for 2.5 MT of LNG over the period of 3.6 years. The second agreement was signed with Qatar Gas for the LNG supply 1.5 mmtpa. For over 20 years, started from 2013. Third agreement was signed with Australia for LNG supply 3.5 mmtpa (Malaysian Insiders, 2013; Kuncinas, 2013).

Other Countries

Japanese companies planned to build new LNG project at the southern coast of Alaska with LNG producing capacity of 20 mtpa. Japan will implement this LNG project by 2018 (The Voice of Russia, 2012). Due to increase in domestic demand, Indonesia is reducing LNG export from traditional LNG Trains. Masela LNG Train 1 and 2 with their storage capacity of (3.4 and 2.7 Bcm) will be started in 2018 (International Energy Agency, 2011). United Kingdom import LNG 10.5 MT in 2012. It has four import terminals with total capacity of 38mtpa (Spomer, 2013). To become LNG exporter Israel need more time. An estimated gas reserves from Tamar and Leviathan field offshore Israel's coast will lead to develop their own LNG project. Estimated gas reserves from both fields are 240 and 450 Bcm (International Energy Agency, 2011).

LNG Trade Worldwide

Before 1990, the total trade of LNG was 53 MT. Only eight importers and eight exporters countries were in the LNG market with 70 LNG vessels in operation. Today the LNG trade scenario is changed and it is increased five times of total supply of 1990. The technology improved with new shipboards, terminals, re-gasification, floating production, large capacity cargo in the new LNG markets worldwide. 25-26 importers and 18 exporting countries were delivered about 240 million tonnes of LNG worldwide in the year 2011-2012 through 365 ships. Number of countries is increasing in the LNG market globally (Walker, 2013; Hill, 2013).

The world largest LNG producing region is the Asia-Pacific. It includes the LNG export/import along the Pacific Rim (including Alaska) and in South Asia (including India). Atlantic Basin includes LNG export/import in Africa, Europe and the Western Hemisphere (Small, 2005). Asia consuming more gas and demand is increasing continuously. According to the published report, LNG imports to Asia in May 2011 were 20.4% more as compared to May 2010. Asia will require 100 million tonnes of LNG by 2020 (LNG World News, 2013). China and Japan are main customers of LNG. Both countries import natural gas from other countries (Tran et al., 2012). New Gulf Coast terminals are expected for more than 70% of United States of America terminal by 2025 (Harrison et al., 2006). Table 2 presents the LNG export and import business by different countries of the world in 2011.

The LNG traditional market all over the world is increasing continuously. The LNG global supply ratio was estimated as 4.5%/year up to 2030. LNG will contribute 25% of global supply in the period of 2010- 2030. Australia will increase LNG supply by 15 Bcf/day from 2014 and accounting for 25% of global LNG production by 2030. Proved gas reserves were estimated about 6609 Tcf in 2010. Shale gas and CBM will account for 63% of North American production by 2030. Growth of shale gas rises the prospect of LNG exports from North America by 2030 (BP Statistical Review, 2012; BP Energy Outlook,

2013). Figure 3 and 4 presents the world LNG supply and demand in 2012.

Table 2: LNG export and import in 2011 (IGU, 2011)

Countries	**LNG export (MT)**	**Countries**	**LNG import (MT)**
Qatar	75.5	Japan	78.8
Malaysia	25	Korea	35.8
Indonesia	21.4	UK	18.6
Australia	19.2	Spain	17.6
Nigeria	18.7	China	12.8
Trinidad	13.9	India	12.7
Algeria	12.6	Taiwan	12.2
Russia	10.5	France	10.7
Oman	7.9	Italy	6.4
Brunei	6.8	USA	5.9
Yemen	6.7	Turkey	4.6
Egypt	6.4	Belgium	4.5
UAE	5.9	Argentina	3.2
Guinea	4.0	Mexico	2.9
Peru	3.8	Chile	2.8
Norway	2.9	Canada	2.4
USA	0.3	Kuwait	2.4
Libya	0.1	Portugal	2.2
Total	24 1 .6	UAE	1.2
		Greece	1.0
		Dom. Rep	0.7
		Thailand	0.7
		Brazil	0.6
		Netherlands	0.6
		Pureto Rico	0.5
		Total	241.8

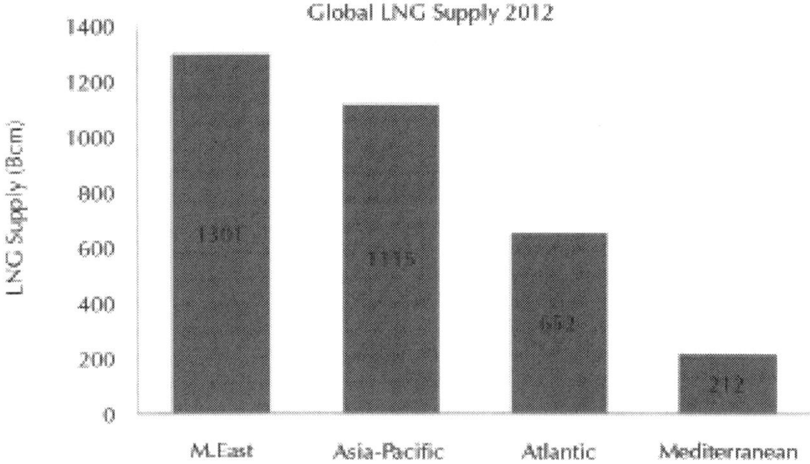

Figure. 3: Global LNG supply 2012 (Stream LNG, 2012).

International Energy Agency (IEA) (2011) estimated that, unconventional gas (production of shale gas) will increase about 25% of the world's gas supply by 2035 (Ernst and Young global energy LNG, 2013). Anadarko Petroleum and Eni agreed for cooperative development program of adjoining offshore areas of Mozambique in December 2012. Good source of gas in place (more than 110 Tcf) has been estimated in the offshore Mozambique. Both companies will plan and construct a general onshore LNG liquefaction/export facility (IHS Global Insights, 2012).

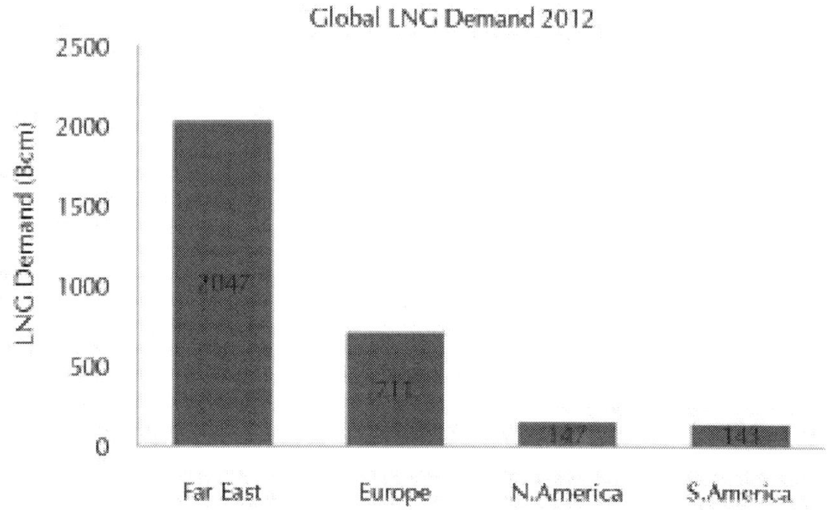

Figure 4: Global LNG demand 2012 (Stream LNG, 2012).

LNG trade should be facilitated in those countries having limited source of natural gas. It is therefore suggested to make the better policies for the customer countries for long time period. In this case the LNG price will be stable for long time.

FUTURE WORLD LNG PROJECTS

Some of the world LNG projects are highlighted in this study. Few of these projects are increasing their current LNG capacity. The first CBM-TO-LNG project by Australia will start the production with capacity 11.6 Bcm in 2014. In Russia, Gazprom planning new LNG project with increasing capacity from 29 Bcm by 2020 to 60 Bcm by 2030. Another project Yamal Peninsula Russia contains 11.7 Tcm of gas reserves the gas production will be expected 185-220 Bcm by 2030. Prelude LNG project Australia with capacity of 4.9 Bcm will complete in 2017. The project is operated by shell. Queensland Curtis LNG Australia led by BG, CNOOC, Tokyo gas with capacity of 11.6 Bcm will start in 2014. Gladstone LNG Australia led by Santos Australia with capacity of 10.6 Bcm will complete in 2015. Gorgon LNG project in Australia led by Chevron, Shell, Exxon Mobil with capacity of 20.4 Bcm will be completed in 2014. Papua New Guinea LNG project with capacity of 9.0 Bcm will be completed in 2014 (International Energy Agency, 2011). The project is operated by Exxon Mobil and Oil search. Donggi Senoro LNG Indonesia led by Mitsubishi, Pertamina KOGAS will start production in 2014. Angola will start LNG production with capacity of 5.2 Mtpa and adding 3.4 million tonnes LNG supply during this year. New import LNG terminals are going to start in Malaysia (LNG re-gasification terminals in Melaka, Pengerang and Tawau), Singapore and Israel (BG Group, 2013; Finance Malaysia Blogspot, 2013; MGA, 2013). Delta Caribe Oriental 1, 2 and 3 LNG project Venezuela with total capacity of 18.2 Bcm will be started in 2015. Kitimat LNG project Canada with capacity of 7.2 Bcm will start in 2015 (International Energy Agency, 2011).

The suggested LNG export projects of North America and East Africa are well conditioned against the LNG projects of Greenfield Australia. This was reported by the analysts at Credit Suisse. They analyzed the price with unit costs averaging less than US$ 2000 per tonne as opposed to Australian average of more than US$ 3000 per tonne (Credit Suisse, 2012). If spot prices remain under the long term agreement therefore, America LNG price will be attractive. Analyst has been assumed that, American spot gas price will be about 5 to 6 US$ per million BTU (DB Research, 2012). For future America based cargos, few of the buyers from the Asia have begun to sign contracts already on the basis of Hennery hub linked prices. Spot prices increase buyer's choices. Flexibility increase in LNG contracts, re-exporting of cargos and other flexibilities to the buyers by the America LNG will contribute to greater linkages between markets and region. This will provide some convergence of regional prices for Asia customers (Macquarie Research, 2012).

According to the Douglas-Westwood, the capital expenditure of new LNG market is expected to grow $169 billion between 2012 and 2016. For the development of LNG import terminal, Asia will be the main investor during 2012-2016. Asia will invest about $31 billion on new LNG import terminals (Dormer, 2012). New discoveries make LNG export projects possible in the new regions of the world.

CONCLUSIONS

Due to decrease of global oil reserves and high oil prices, the strongest LNG demand will be increased globally. The world energy source LNG play an important role in the power generation due to its flexible prices. The technology is growing gradually with increasing number of LNG consuming countries all over the world. LNG is the only source of energy that can transport and provide in the cheap price from long distance to fulfill the demand by the countries of the world.

In the near future the most rapidly growing LNG producer in the world is Australia. After Qatar, the world second largest LNG exporter will be the Australia. Australia will increase LNG supply by 15 Bcf/day from 2014 and accounting for 25% of world LNG production by 2030.

Global LNG production forecast will be reached 540 Bcm by 2020 and LNG trade will be reached 425 Mtpa by 2025.

United States of America and China will minimize their LNG demand in the near future due to their domestic production of shale gas.

ACKNOWLEDGMENTS

We would like to acknowledge the Universiti Teknologi PETRONAS for supporting this research study.

REFERENCES

1. Abdullah, M., 1998. Trends for future gas supply planning for gas utilization projects in Malaysia. Proceedings of the SPE Asia Pacific Oil and Gas Conference and Exhibition. Perth, Australia.
2. Barclay, M. and T. Shukri, 2005. Enhanced Single Mixed Refrigerant Process for Stranded Gas Liquefaction. Retrieved from: http://www.foster wheelerag.ch/publications/tech_papers/.
3. BG Group, 2013. Global LNG Market 2012-2013. Retrieved from: http://www.bg-group.com. (Accessed on: May 9, 2013).

4. BP Energy Outlook, 2013. BP Energy Outlook 2030. pp: 1-86. Retrieved from: www.bp.com/en/global/ .bp/.energy-2013/ energy-outlook-2030.html.

5. BP Statistical Review, 2012. BP Statistical Review of World Energy June 2012. pp: 1-48. Retrieved from: www.bp.com/.../ bp.../statistical_energy_ review.../statistical_review_of_worl.

6. Choi, M.S., 2010. LNG for petroleum engineers. SPE Proj. Fac. Const., 6(4): 255-263.

7. Credit Suisse, 2012. Credit Suisse Global Equity Research Global LNG Sector. Retrieved from: https://www.credit-suisse.com. (Accessed on: June 7, 2012).

8. DB Research, 2012. Deutsche Bank Markets Research, Global LNG. Google Search. Retrieved from: http://www. dbresearch. com/. (Accessed on: September 17, 2012).

9. Dormer, M., 2012. LNG Import and Export Growth to be Driven by Activity in the Pacific Basin. Douglas, Westwood, pp: 2016-2018. Retrieved from: www.douglas-westwood.com.

10. Ernst and Young Global Energy LNG, 2013. Will New Demand and New Supply Mean New Pricing? Retrieved from: www. ey.com/oilandgas. (Accessed on: April 6, 2013).

11. Finance Malaysia BlogSpot, 2013. IPO: Gas Malaysia. Retrieved from: http://financemalaysia.blogspot. com/2012/06/ipo-gas-malaysia.html. (Accessed on: June 3, 2013).

12. GIIGNLG, 2012. International Group of Liquefied Natural Gas Importers. Net Global LNG Trade Slides in 2012. Retrieved from: www.ogj.com. (Accessed on: June 10, 2013).

13. Harrison, M.C., P. Campbell, F.M. Coelho, C. Grecco, C. Ikeocha, X.T. Wang, M.J. Economides et al., 2006. Natural gas: Overcoming limitations toward energy supremacy. Proceeding of the SPE Annual Technical Conference and Exhibition. San Antonio, Texas, pp: 1-8.

14. Hill, S., 2013. LNG-Are The Dynamics Changing? BG Group. Retrieved from: files.the-group.net/ library/bggroup/files/ doc_392.pdf.

15. IGU (International Gas Union), 2011. Waterborne LNG Reports, US DOE, PFC Energy. World LNG Report 2011, Retrieved from:

www.igu.org/gasknowhow/.../igu.../LNG%20Report%202011.pdf.

16. IHS Global Insights, 2012. Anadarkes and Eni Sign Gas Deal With Mozambique Government Awards FEEO Contracts. (Accessed on: December 21, 2012).
17. International Energy Agency (IEA), 2011. Mediumterm Oil and Gas Markets 2011. Retrieved from: http://www.iea.org. (Accessed on: May 5, 2013).
18. Koh, Q., 2013. Malaysia Aims to Become LNG Trading Hub Through New $1.3B Terminal. Retrieved from: http://www.rigzone.com. (Accessed on: March 15, 2013).
19. Kuncinas, P., 2013. Malaysia: Push for Liquefied Natural Gas. Borneo Post Online. Retrieved from: http://www.the borneopost.com. (Accessed on: March 14, 2013).
20. LNG World News, 2013. The Industry's LNG News Terminal. Retrieved from: http://www. lngworldnews.com. (Accessed on: May 5, 2013).
21. Macquarie Research, 2012. Macquarie Equity Research. Global LNG Outlook, (Accessed on: September 10, 2012).
22. Malaysian Insiders, 2013. Malaysia Imports LNG in Struggle to Eliminate Subsidies. Retrieved from: http://www.themalaysianinsider.com. (Accessed on: February 23, 2013).
23. MEO Australia, 2013. Liquified Natural Gas. Retrieved from: http://www.meoaustralia.com. (Accessed on: Feburary 15, 2013).
24. MGA (Malaysian Gas Association), 2013. The Gas Industry in Malaysia. Retrieved from: http://www.malaysiangas.com. (Accessed on: June 3, 2013).
25. Mokhatab, S. and M.J. Economides, 2006. Onshore LNG production process selection. Proceeding of the SPE Annual Technical Conference and Exhibition. San Antonio, Texas, pp: 1-11.
26. OECD/IEA (International Energy Agency), 2013. Developing a Natural Gas Trading Hub in Asia: Obstacles and Opportunities. Retrieved from: http://www.iea.org. (Accessed on: May 5, 2013).
27. Oil and Gas Journal (OGJ), 2013. International Group of LNG Importers. The LNG Industry in 2012. Retrieved from: http://www.ogj.com. (Accessed on: May 7, 2013).

28. Petronas Malaysia MLNG, 2013. Retrieved from: http://www.mlng.com.my. (Accessed on: March 1, 2013).
29. Qatar LNG to Asia, 2012. Qatar Looking to Export US Shale Based LNG to Asia. Retrieved from: http://www.theaustralian.com.au/business/ miningenergy/. (Accessed on: June 10, 2013).
30. Small, D., 2005. The Global LNG Industry-Changed Market Dynamics. Proceeding of the 18th World Petroleum Congress. Johannesburg, South Africa, Sept. 25-29, pp: 1-11.
31. Spomer, B., 2013. Natural Gas Challenges and Drivers for the Future. BG Group. Retrieved from: http://www.bg- group.com/ InvestorRelations/ Presentations/ Documents/ BG Group-IPweek_BetsySpomer-Feb-2013.pdf.
32. Stream LNG, 2012. LNG Trade and the Impact of Technology. Retrieved from: http://www. streamlng.com. (Accessed on: April 6, 2013).
33. The Edge Malaysia, 2013. PETRONAS Gas: Pengerang LNG Re-Gas Plant Confirmed in the Edge. Retrieved from: www.theedgemalaysia.com. (Accessed on: March 1, 2013).
34. The Voice of Russia, 2012. Russia to Strengthen its Position on LNG Market 2012. Retrieved from: voiceofrussia.com. (Accessed on: June 10, 2013).
35. Tran, C., B. Cassidy, M. Pierce, X. Wang and M.J. Economides, 2012. International LNG prospects 2011 and beyond. Proceeding of the International Petroleum Technology Conference. Bangkok, Thailand, pp: 1-12.
36. US EIA Qatar (US Energy Information Administration), 2013a. Qatar Updated January 2013. Retrieved from: http://www.eia.gov/ dnav/ng/ng. (Accessed on: June 10, 2013).
37. US Energy Information Administration (US EIA), 2013b. US Natural Gas Export by Country. Retrieved from: http://www.eia.gov/dnav/ng/ng. (Accessed on: June 10, 2013).
38. Walker, A., 2013. Global LNG Market-a Look Back and a Look Forward. Retrieved from: http://www. bg-group.com. (Accessed on: May 9, 2013).

Chapter 9

Naturally Derived Biomaterials: Preparation and Application

Tran Le Bao Ha[1], To Minh Quan[1], Doan Nguyen Vu[1], and Do Minh Si[1]

[1]University of Science, Vietnam National University – Hochiminh city, Vietnam

INTRODUCTION

The success of any implant depends so much on the biomaterial used. Naturally derived biomaterials have been demonstrated to show several advantages compared to synthetic biomaterials. These are biocompatibility, biodegradability and remodeling. Therefore, these biomaterials are usually applied in the repair or replacement of damaged human tissues and organs. The aim of this chapter is to provide a brief knowledge of naturally derived biomaterials as well as methods of preparation and application of them.

Biomaterials can be classified into two main groups: synthetic and natural biomaterials. Synthetic biomaterials are classified as: metals, ceramics, nonbiodegradable polymers, biodegradable polymers... Some synthetic biomaterials are commercialized and applied in clinical treatment such as metal hip, Dacron, plastic intraocular lens... However, synthetic biomaterials have some disadvantages, including their structure and composition is not similar to native tissues/organs, their biocompatibility and their ability to induce tissue remodeling are low. Thus, other biomaterials have been developed that can overcome the disadvantages of synthetic biomaterials. Today, naturally derived biomaterials have been attracting scientist's interest all over the world. Naturally derived biomaterial can be classified into many groups including protein-based biomaterials (collagen, gelatin, silk...), polysaccharide-based biomaterials (cellulose, chitin/chitosan, glucose...) and decellularized tissue-derived biomaterials (decellularized heart valves, blood vessels, liver...). Protein and polysaccharide-based biomaterials can be prepared by two distinct ways. Protein and polysaccharide from living organisms are dissolved by solvents or enzymes. Then, they are precipitated and reconstituted into fibrils. The second way to prepare protein and polysaccharide is removing other elements of living organisms by solvents or enzymes. Decellularized biomaterials are created by eliminating all cells from native tissues/organs. Physical, chemical and enzymatic approaches are combined to make the effective decellularization protocol.

Because of their advantages, naturally derived biomaterials are usually applied to replace or restore structure and function of damaged tissues/organs. They have ability to adequately support cell adhesion, migration, proliferation and differentiation. In particular, when implanted into a defective area, naturally derived biomaterials can enhance the attachment and migration of cells from the surrounding environment, therefore, induce extracellular matrix formation and promote tissue repair. Some biomaterials are used to acting as drug delivery system and medical devices such as surgical sutures. The silk fiber produced by silkworm or spider has been used as a surgical suture for a long time due to its biodegradable and non-antigenic protein. These silk fibroin nanoparticles are the globules with a fine crystallinity that may offer various possibilities for surface modification and covalent drug attachment. Furthermore, some biomaterials are used to produce environmental friendliness of packaging (such as resorbable chitosan packing) and other products. Some commercial products were made from naturally derived biomaterial such as SIS, Matrigel, Alloderm... In this chapter, we focus on a brief knowledge as well as the methods of preparation and application of naturally derived biomaterials in our researches.

NATURALLY DERIVED BIOMATERIALS: PREPARATION AND APPLICATION

Protein

Structure

Collagen is the most abundant protein of connective tissues in all animals. Now, at least sixteen types of collagen have been identified, in which 80-90% of the collagen is types I, II and III. Collagen is secreted by not only fibroblasts but also epithelial cells [1].

The basic structural unit of collagen is a triple helix. Most collagen is fibrillar because of pack of collagen molecules type I, II, III. Contrast, collagen IV forms a two dimensional network which is unique to the basement membranes [1]. Basement membranes have been performed a number of mechanical and biological functions. They provide physical support for tissue because of their tensile strength. They also influence cell proliferation, adhesion, migration, differentiation, polarization, and are thus implicated in biological processes such as development, tissue maintenance, regeneration, and repair, and in various pathological processes such as tumor growth and metastasis [2].

The basement membranes composition varies from one tissue to another. In general, the major constituents of all basement membranes are collagen IV, laminins, nidogen/entactin, and proteoglycans. The functional diversity of basement membranes arises from the molecular diversity of their components, particularly the different collagen IV and laminin isoforms [2].

Preparation

Collagen can be obtained from various sources, in which amniotic membrane (AM) is an attractive source. AM is a thin membrane surrounding the fetus which is filled with amniotic fluid.

The AM consists of an epithelial monolayer, a thick basement membrane, a compact layer, a fibroblast layer and a spongy layer [3]. The innermost layer, nearest to the fetus, is monolayer of epithelial cells anchored on the basement membrane. The collagen component of basement membrane of AM includes types III, IV, V, VII, XVII which similar morphological and ultra-structural basement membrane of skin. Therefore, basement membrane of AM is often used to create skin equivalents. Besides, AM has outstanding properties such as anti-inflammatory, anti-bacterial, anti-fibrosis, anti-scaring as well as low immunogenicity and reasonable mechanical features [3].

AM can be used either with amniotic epithelium (intact) or without it (denuded), fresh or preserved. To remove the amniotic epithelium, the AM is incubated in trypsin-EDTA at 37°C in 30 min and the cells are gently scraped while maintaining the intact basement membrane. H&E staining was performed to confirm removing the amniotic epithelium. Then, the basement membrane can be preserved by drying or glycerol - cryopreservation after -sterilization [3, 4, 5].

Application

Collagen is commonly used in biomedical applications. The basement membrane of the AM is a typical example. The extracellular matrix components of the basement membrane of the AM are native scaffolds for cell seeding in tissue engineering. AM has been applied in tissue engineering related to eye, skin, cartilage, nerve, especially cancer [3, 6, 7].

To use AM in creating skin equivalent, AM must be removed the epithelium. Keratinocytes are seeded onto basement membrane which is denuded AM. Briefly, keratinocytes were isolated from intact skin samples by incubation in trypsin-EDTA at 4^0C in 18 hours and detached mechanically. Keratinocytes were cultured in serum free medium in 7 days. Medium was chanced every two days [8]. After 3rd passage, cells were subcultured onto AM basement membrane which spreaded on bottom of the insert dishes. The cells were maintained in culture for 7 days when the cells reached confluent. Air-lifting was performed to induce cell differentiation. After 7 days, the cells formed multi-layers on the AM basement membrane. Cultured keratinocyte sheets were grafted on patients who were defected skin because of injury or burn. The result showed that, the advantages of cultured keratinocyte sheet autograft: the possibility of the grafting area multiplication (50 times after 3 weeks), the diminution of scaring, the relief of pain, the low infection risk, the same effect in compare to the split - thinness autograft.

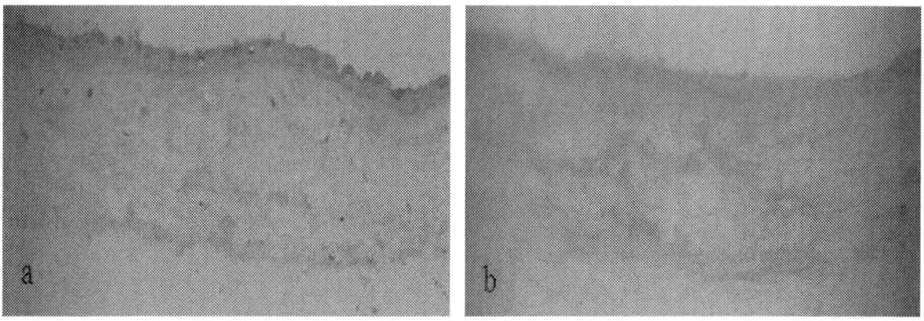

Figure 1: Amniotic membrane with epithelium: intact (a) and denuded (b).

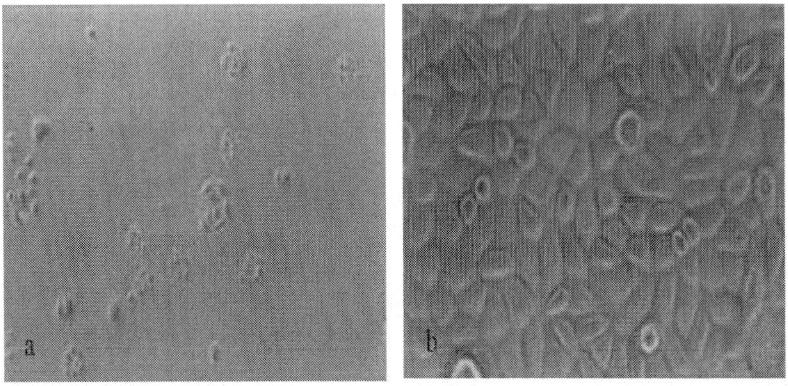

Figure 2: Keratinocytes formed colonies (a) and monolayer onto culture dishes (b) (200X).

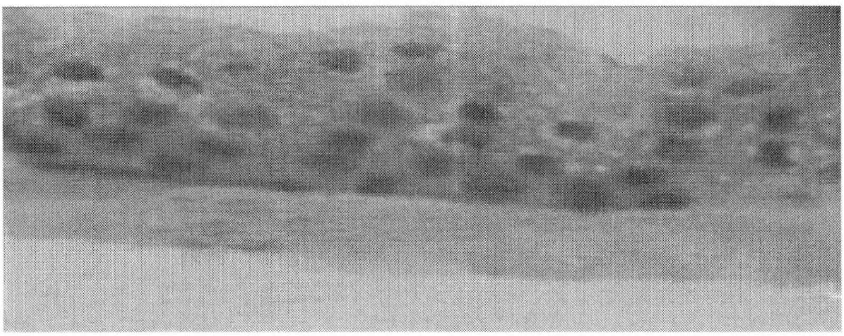

Figure 3: Result of HE staining of the cell sheet, the cells were exposed to the air in seven days. The cells formed 4 – 5 layers onto AM basement membrane (400X).

Gelatin-Alginate

Structure

Gelatin is obtained by controlling the hydrolysis of collagen, a fibrous insoluble protein which is widely found in nature and is the major component of skin, bone and connective tissue. Characteristic features of gelatin are the high content of the amino acids such as glycine, proline and hydroxyproline. Structually, gelatin molecules contain repeating sequences of glycine-X-Y triplets, where X and Y are frequently proline and hydroxyproline. These sequences are responsible for the triple helical structure of gelatin and its ability to form gels where helical regions form in the gelatin protein chains immobilizing water [9].

Alginate was first discovered by Edward Stanford in 1883. Since being commercialized in 1927, alginate has now expanded to about 50.000 tonnes per year worldwide; 30% of this tonnage is devoted to the food industry, the rest being used in industrial, pharmaceutical and dental applications [10]. The function of alginates in algae is primarily skeletal, with the gel located in the cell wall and intercellular matrix conferring the strength and flexibility necessary to withstand the force of water in which the seaweed grows [11].

Alginate is a hydrophilic polysaccharide extracted from marine brown algae such as Laminaria hyperborea or soil bacteria such as Azobacter vinelandii and composed of 1,4-linked -D-mannuronic acid (M) residues and 1,4-linked -L-guluronic acid (G) in varying proportions, displaying carboxylic acid functionality at the C5 residue. The alginates have broad distributions of molecular weights of 10-1000 kDa depending on source and processing. The relative amount and sequential distribution of homogeneous M-M segments (M-blocks), homogeneous G-G segments (G-blocks) and alternating M-G segments (MG-blocks), which represent the primary structure of alginate, depend on the producing species, and for marine sources, on seasonal and geographical variations.

Preparation

Cross-linked gelatin/alginate was made in two steps. Briefly, 1wt % (w/w) aqueous solution of gelatin and sodium alginate, respectively, was dissolved in double distilled water at 500C for 3 h. Each solution with certain mixing ratios of gelatin and sodium alginate (8G:2A) was stirred for 30min at room temperature, frozen to -700C for 40 h. This soluble sponge was cross-linked with EDC by immersing the soluble sponge in 90% (w/v) aqueous acetone containing 0.3% EDC for 24 h at room temperature, while shaking slowly.

The sponges consisting of gelatin from hydrolysis of collagen and alginate-a polysaccharide from Phaeophyta were established by using EDC as a crosslinking agent. The sponges were combined with some natural substances traditional used for burn treatment such as tamanu oil (from nuts of Calophyllum inophyllum); cajeputi oil (from leaves of Meulaleuca leucadendron); madecassol (from extract of Centella asiatica); turmeric and python fat. Data obtained from testing on mice showed that the coordinated sponges have rather good ability on preventing infection and promoting wound healing compared with control. The sponges combined with mixture of cajeputi oil and madecassol have the best potential for burn treatment.

Figure 4. Gelatin-Alginate. Block (a), SEM (b).

Application

Gelatin has been used in medicine as plasma expander, wound dressing, adhesive, and absorbent pad for surgical use. While collagen, also known to have wide biomedical applications, expresses antigenicity in physiological condition, gelatin is known to have no such antigenicity. Recently, gelatin has shown to exhibit activation of macrophages and high hemostatic effect. Finally, gelatin is practically more convenient than collagen because a concentrated collagen solution is extremely difficult to prepare from the native collagen, and furthermore gelatin is far more economical than the collagen.

Alginate is an effective natural disintegrant, tablet binder and offers an attractive alternative for sustained-release systems. It offers advantages over synthetic polymers as it forms hydrogels under relatively mild pH and temperature and is generally regarded as non-toxic, biocompatible, biodegradable, less expensive and abundantly available in nature; in addition, alginate meets the important requirement of being amenable to sterilization and storage. All these advantages make alginates very useful materials for biomedical applications, especially for controlled delivery of drugs and other biologically active compounds and for the encapsulation of cells. Calcium alginate is a natural haemostat, so alginate based dressings are indicated for bleeding wounds. The gel forming property of alginate helps in removing the dressing without much trauma [12].

Alginate has been much used in medical applications such as wound dressings, scaffolds for hepatocyte culture and surgical or dental impression materials. Alginates are also known to be broken down to simpler glucose type residues and can be totally absorbed.

Alginate has been successfully used as a matrix for the entrapment and/or delivery of biological agents, such as drugs and proteins. In particular proteins can be loaded and released by alginate matrices without loss of their biological activity because of the relatively mild gelation process of alginate. In pharmaceutical formulations, the alginate gel can be prepared prior to use, or it can spontaneously form in situ in physiological fluids, by low pH and/or calcium ions naturally present in the site of administration [13]. Alternatively, the gelling agent can be added either as a part of the formulation or separately administered. The microencapsulation technique has been specifically developed for the oral delivery of proteins, as they are quickly denaturated and degraded in the hostile environment of the stomach.

Among the possible applications of alginate, one of the most promising is for cell immobilization. Alginate gel allows cell suspension to be cultivated in several types of bioreactors to achieve high cell densities [14]. In cell immobilization applications, the main drawback of alginate matrix gels is represented by their high density of network, which limits the cell growth; moreover, cell anchorage, a strict requirement for survival, is limited on alginate gels, because of its hydrophilic nature. PEG copolymers are used to improve their biocompatibility [15].

Silk

Silks are generally defined as protein polymers that are spun into fibers by Lepidoptera larvae such as silkworms, spiders, scorpions, mites and flies [16]. Silks are fibrous proteins synthesized in specialized epithelial cells that line glands in these organisms. Silk fibroin polymers consist of repetitive protein sequences and provide structural roles in cocoon formation, nest building, traps, web formation, safety lines and egg protection. The most extensively used silk for various applications are those from silkworm silk; Bombyx mori and spider silk; Nephila clavipes.

The domesticated silkworm (B. mori) silk fibroin fibers are about 10–25 μm in diameter. Each fiber consists of core protein covered by a coating protein (sericin) that glues core fibers together. The core protein consists of three chains: heavy chain, light chain and a glycoprotein, P25. The light chain (26 kDa) and heavy chain (390 kDa) which are present in a 1:1 ratio and linked by a single disulfide bond. The disulfide linkage between the Cys-c20 (20th residue from the carboxyl terminus) of the heavy chain and Cys-172 of the light chain holds the fibroin together and a 25 kDa glycoprotein, named P25, is non-covalently linked to these proteins [17]. Light chain is necessary for the secretion of protein from the silk glands. Heavy chain is fiber forming protein and its structure determines properties of silk fiber [18]. Heavy chain is commonly referred as fibroin protein. These proteins are coated with a family of hydrophilic proteins called sericins (20–310 kDa). Silk proteins are particularly promising for these needs due to their unique combination of biocompatibility, biodegradability, self-assembly, mechanical stability, controllable structure and morphology.

Spiders are look like insects and come in to the category of Arthropoda, but they belong to a completely different class of animals, called Arachnida. Spider silk is the result of 400 million years of evolution. Spiders use silk for variety of functions including reproduction as cocoon capsular structures, lines for prey capture, lifeline support (dragline), web construction and adhesion. Spider silk is a biopolymer fiber. Its composition is a mix of an amorphous polymer (which makes the fiber elastic), and chains of two of the simplest proteins (which give it toughness). Out of 20 amino acids, only Glycine and Alanine serve as a primary constituent of silk. The Dragline silk of orbweb spider seems to be most studied in the scientific research. The protein in dragline silk is fibroin (Mass of 200,000-300,000 Daltons) which is a combination of the proteins spidroin 1 (Alanine-rich) and spidroin 2 (Glycine-rich), the exact composition of these proteins depends on species. Fibroin consists of approximately 40% Glycine and 25% Alanine as the major amino acids. The remaining components are mostly glutamine, serine, leucine, valine, proline, tyrosine and arginine [19]. Nephila clavipes can produce seven types of silk from seven different silk glands as shown in depending on needs and environmental conditions [20]. The superior mechanical properties of dragline spider silks can be used as a template for developing specific structures for various biomaterial needs. Spider silks have not been commercialized in fashion as silkworm silk due to the lack of domestication and lower productivity of spiders.

Preparation

Nanotechnology is becoming a key technology and capable of application in all fields of science and technology. In particular, nanoparticle delivery system significantly improved pharmaceutical treatment of many incurable diseases which require complex treatment regimens, as well as, patients must take multiple medications and need time long-term drug use. Silk Protein is the ideal material for this purpose, because they have many unique features such as highly biocompatible and biodegradable ability, self-restructuring, mechanical stability, easy control and adjustment of the object's structure and shape.

Figure 5. Nanofibroin particles.

The cocoon shell of silkworm Bombyx mori was degummed in boiling solution of 0,5% Na_2CO_3 in 700C for 35 min. Then degummed fiber was dissolved in a mixed solution of calcium chloride, ethanol, and water ($CaCl_2/C_2H_5OH/H_2O$: 1:2:8 mole ratio), at 800C. After the silk fibroin–salts solution was centrifuged at 5000 rpm for 10 min, the supernatant was dialyzed continuously for 72 h against running pure water to remove $CaCl_2$, smaller molecules, and some impurities. The resulting liquid silk fibroin was stored at 40C and used in the following experiments for the preparation of silk fibroin nanoparticles. Spider silk proteins form nanoparticles upon salting out with potassium phosphate. Milk-like silk protein particles were formed at once and suspended in the mixture comprising water and organic solvent. These protein particles were water insoluble and went down slowly due to the gathering of microparticles. The precipitates of silk protein nanoparticles were collected and purified from the mixture by repeated centrifugation at 20,000 rpm to separate these particles from the solvent. After the research, we have obtained nanoparticles (500 nm-2000 nm) from silk protein can load and delivery of proteins in vitro.

Application

The silk bio-polymer is used in tissue regeneration for treating burn victims and as matrix of wound healing. The silk fibroin peptides are used in cosmetics due to their glossy, flexible, elastic coating power, easy spreading and adhesion characters [21]. Silk powder is touted and relieves from sunburns, due to crystalline structure it reflects UV radiation and as demulcent it acts as protective buffer between skin and environment. The lower micron silk powder is added with hair and massage oils and water dispersible finer grade silk powder is an ingredient of liquid cosmetic preparations.

The silk is used to fight edema, cystitis, impotence, adenosine augmentation therapy, epididymitis and cancer [22]. Silk protein derivative, Serratio peptidase is used as anti-inflammatory, anti-tumefacient for treating acute sinusitis, tonsiloctomy, oral surgery, tooth filling, cleaning and extractions. The silk fibroin is a useful dressing material with the property of non-cytotoxic to the tissues and also in veterinary medication.

Since long, silk fiber is being used as surgical sutures as it does not cause inflammatory reactions and is absorbed after wounds heal. Other promising medical applications are as biodegradable micro tubes for repair of blood vessels and as molded inserts for bone, cartilage and teeth reconstruction [23, 24,25]. In biomedical and bioengineered field, the use of natural fibre mixed with biodegradable and bio-resorbable polymers can produce joints and bone fixtures to alleviate pain for patients.

Drug delivery is a rapidly developing field in biomedical research. It is interdisciplinary and requires expertise in biotechnology, pharmacology, microbiology, biochemistry, polymer chemistry and materials engineering. Advantages of using such systems include maintenance of drug levels within desired range, fewer administrations, optimal use of the drug, and better patient compliance. The material used for drug delivery should be biocompatible, chemically inert, easily processable and physically and mechanically stable. Biopolymers are of great interest for this kind of application. Silk and silk-like variants are used by some scientists as carriers for drug delivery. Their biocompatibility and ability to form hydrogels in situ makes them attractive candidates for the localized, controlled delivery of therapeutic agents. Their ability to incorporate drugs at room temperature, by simple mixing, and without the use of toxic or denaturing solvents makes them attractive for the delivery of protein or DNA-based therapies [26].

Future applications of silk biomaterials include new generation soft contact lenses that enable greater oxygen permeability, artificial corneas, skin grafts and epilepsy drug permeable devices.

Fibrin

Structure

The mechanism of fibrin fomation is elucidated primarily from the thrombin-mediated cleavage of fibrinogen. Fibrinogen, the principal protein of blood clotting, is a 340 kDa trinolar protein which presents at high concentration in blood plasma (2 – 4 mg/ml, 6 – 12 μM). Fibrinogen molecule consists of three different pairs of polypeptide chains (A , B and) cross-linked to each other by 29 disulfide bridges (Fig. 1). The amine-termini (N-termini) of six polypeptide chains are converged in the central of fibrinogen molecule called the E domain. The carboxy-termini (C-termini) of the B chain and chain comprise of the distal D domain. The C-termini of A chains which are known as globular, depart from the D domain and fold into a conformation that stretches back toward the E domain of fibrinogen [27].

Fibrinogen plays as precursor protein of fibrin in blood clotting. The conversion of fibrinogen to fibrin occurs in 3 ordered steps. In the intinial step, thrombin binds to the central E domain of fibrinogen and slipts off the fibrinopeptides A – FpA (16 amino acid residues) and B – FpB (14 amino acid residues) from N-termini of A and B chains, respectively, whereas the chains remain unaltered. The cleavage of FpA and FpB results in exposure of "A" and "B" binding sites. Then, the self-assembly proccess will spontaneously occur. The "A" and "B" sites will interact with complementary sites ("a" and "b" sites located in the and chain) at the D domain of other fibrinogen molecules, which results in new fibrin monomers. The fibrin monomers are bound to each other non-covalently (Fig. 2) and assemble in a half-staggered manner into two-stranded protofibrils which continue to aggregate laterally to form fibers branching into a three-dimensional network of fibrin [28, 29]. Finally, fibrin cross-linking is activated by Factor XIII (FXIII) in order to improve the strong and elastic properties, additionally, avoid fibrinolysis in solution [30, 31].

Fabrication of Fibrin Gel

According to the usage purposes, some methods have been applied to fabricate fibrin gels. Fibrin gels can be conducted either from the separating components including thrombin, fibrinogen and CaCl2, or from serum of patients, which will be mentioned in two following methods, respectively. For manipulation of fibrin gel from separating commercial components, fibrin gels were prepared by combining fibrinogen, NaCl, thrombin, CaCl2. This complex is also supplemented with aprotinin in order to proving a stable fibrin structure and prevents postoperative bleeding. The contents are allowed to gel for 1 hr in standard culture conditions [32]. Furthermore, fibrin gel physical properties can be manipulated by adjusting the fibrinogen and $CaCl_2$ concentration [33, 34], or using different cross linking agents such as enzymes or UV radiation [35]. In terms of autologous fibrin glue, the patient blood is havested and prepared 3 to 4 days before surgery. The plasma is separated from red blood cells by allowing the blood tube to stand vertically for at least 2 hours or centrifuged at 4000 rmp for 5 minutes. The fibrin gel preparation is created by combining plasma with commercial thrombin and calcium at appropriate concentration [36]. Autologous plasma fibrin gel not only shows an excellent hemostatic agent, but also helps eliminate the risk of viral transmission associated using donor plasma.

Fibrin Gel Applications

Numerous studies have exploited fibrin function as heamostatic plug, scaffold for cell proliferation and migration, and wound healing, which suggest fibrin potential applications in medical and tissue engineering. Fibrin glue or fibrin sealant is a formulation of fibrinogen and thrombin at very high amounts cobined with calcium and FXIII, used as an adjunct to hemostasis in patients undergoing surgery. Commercial products of fibrin sealant such as Tisseel (Immuno, Vienna, Austria), Beriplast (Behringwerke AG, Marburg/Lahn, FRG), and Biocol (CRTS, Lille, France) have been extensively used in clinical. In addition to fibrin'role in heamostatic, fibrin is also indicated as biological scaffold for cell proliferation, migration and differentiation applied in various tissue engineering. Natural fibrin matrix consists of sites for cellular binding, and has been shown to have excellent effects in cell culture and accelabrate tissue regeneration. In 2000, Ye et al. fabricated and investigated the three-dimensional fibrin scaffold in cardiovascular tissue engineering. In this research, human myofibroblasts (MFBs) from the ascending aorta were cultured in fibrin gel solution. Consequently, the cell growth, high collagen secretation and tissue development were determined. Besides, toxic degradation or inflammatory reactions was not detected in the fibrin gels [37]. In 2003, W. Bensa.ıd and colleagues conducted a research in which they use fibrin glue as a delivery system for human MSCs (HMSCs). The result confirmed a good good HMSCs spreading and proliferation in the fibrin scaffold. Besides, the HMSCs migration out of the fibrin scaffold and appearance of calcium carbonate from the differentiation of HMSCs when implanted in vivo suggest that fibrin gel is a promising delivery system for HMSCs toward bone healing application [38]. Fibrin glue also performs its role in the application of skin grafts to burned areas. Using fibrin glue instead of sutures or pressure dressings in the immediate postoperative period enhances healing, and minimizes scarring [39]. One of commercial fibrin sealant products used for burn treatment is ARTISS fibrin sealant (Baxter International Inc., USA). ARTISS fibrin sealant is indicated to adhere autologous skin grafts to surgically prepared wound beds resulting from burns, for both adults and pedipatients.

POLYSACCHARIDE

Cellulose

Structure

Cellulose is the most abundant polymer on Earth, which makes it also the most common organic compound. Annual cellulose synthesis by plants is close to 10^{12} tons. Plants contain approximately 33% cellulose whereas wood contains around 50% and cotton contains 90%. Most of the cellulose is utilised as a raw material in paper production. This equates to approximately 10^8 tons of pulp produced annually. From this, only 4 million tons are used for further chemical processing annually. It is quite clear from these values that only a very small fraction of cellulose is used for the production of commodity materials and chemicals [40]. Cellulose, a linear polysaccharide of up to 15,000 D-glucose residues linked by β-(1→4)-glycosidic bonds, is biocompatible and has excellent thermal, mechanical properties. It is considered easily biodegradable, thus less contaminating to the environment.

Cellulose is regarded as a semi-flexible polymer. The relative stiffness and rigidity of the cellulose molecule is mainly due to the intramolecular hydrogen bonding. This property is reflected in its high viscosity in solution, a high tendency to crystallise, and its ability to form fibrillar strands. The chain stiffness property is further favoured by the β-glucosidic linkage that bestows the linear form of the chain. The chair conformation of the pyranose ring also contributes to chain stiffness. This is in contrast to the α-glucosidic bonds of starch [41].

Plants are an attractive cellulose source primarily because they are abundant and there is a preexisting infrastructure in the textile industries for harvesting, retting/pulping (i.e. to treat and isolate micron sized cellulose particles), and product processing. Tunicates are the only animals known to produce cellulose microfibrils. Tunicates are a family of sea animals that have a mantle consisting of cellulose microfibrils embedded in a protein matrix. It is this thick leathery mantle in their mature phase that is used as a source of cellulose microfibrils. Most research has used a class of Tunicates that are commonly known as "sea squirts" (*Ascidiacea*), marine invertebrate filter feeders. Several algae species such as green, gray, red, yellow-green... produce cellulose microfibrils in the cell wall. There are considerable differences in cellulose microfibril structure between the various algae species caused by differences in the biosynthesis process. Most cellulose microfibril researchers have used various species of green algae. Bacterial cellulose (BC) is a glucose polymer produced through bacterial fermentation. This macromolecular polymer features the same molecular formula and properties of natural cellulose. A fiber bundle of 40 to 60 nm thick is formed by micro-fibers with a diameter range of 3 to 4 nm. These bundles aggregate randomly to produce a developed structure forming a typical type of nanobiomaterial [42].

Cellulose derivatives and composites offer an excellent biocompatibility, and are considered as promising materials for biochemical engineering for economic and scientific reasons.

- Oxidized cellulose (oxycellulose) is cellulose in which some of the terminal primary alcohol groups of the glucose residues have been converted to carboxyl groups. Therefore, the product is possibly a synthetic polyanhydrocellobiuronide and that contain 25% carboxyl groups are too brittle and too readily soluble to be of use. Those products that have lower carboxyl contents are the most desirable [43].
- Purified microcrystalline cellulose (MCC) is partially depolymerized cellulose prepared by treating α-cellulose, obtained as a pulp from fibrous plant material, with mineral acids. Silicified MCC (SMCC) is manufactured by codrying a suspension of MCC particles and colloidal silicon dioxide such that the dried finished product contains 2% colloidal silicon dioxide. SMCC shows higher bulk density than the common types of MCC. Also, tensile strength of compacts of SMCC is greater than that of the respective MCC and it is most probably a consequence of intersurface interactions of silicon dioxide and MCC [44].
- The esterification can be considered as a typical equilibrium reaction in which an alcohol and acid react to form ester and water. Cellulose is esterified with certain acids such as acetic acid, nitric acid, sulfuric acid, and phosphoric acid. A prerequisite is that the acid used can bring about a strong swelling thus penetrating throughout the cellulose structure. Cellulose acetate phthalate is a partial acetate ester of cellulose that has been reacted with phthalic anhydride. One carboxyl of the phthalic acid is esterified with the cellulose acetate. The finished product contains about 20% acetyl groups and about 35% phthalyl groups [45].

Preparation

The treatments for wood and plants involve the complete or partial removal of matrix materials (hemicellulose, lignin, etc.) and the isolation of individual complete fibers. Fortunicate the treatment involves the isolation of the mantel from the animal and the isolation of individual cellulose fibrils with the removal of the protein matrix. Treatments for algal cellulose sources typically involve culturing methods, and then purifying steps for removal of algal wall matrix material. Bacterial cellulose treatments focus on culturing methods for cellulose microfibrillar growth and then washing to remove the bacteria and other media.

The general processing of engineered BC materials can be considered to occur in four main stages: (1) BC culturing, (2) pellicle management, (3) water removal, and (4) chemical modification. For stage 1, the biosynthesis of BC occurs in culture solutions, generally in a bioreactor, in which bacteria secrete cellulose microfibrils, producing an interwoven web of fibrils that is a hydrogel. The hydrogels are composed of entangled cellulose microfibrils formed from the random motion of the bacteria, contain upwards of 99% water, and are called pellicles. For stage 2, pellicle management refers to any process imparted on the pellicle up until the point of water removal. To remove the bacteria from the pellicles, the pellicles are washed by boiling in a low concentration (2%) NaOH bath for several hours, then it is rinsed under running water for several days. Additional NaOH and NaClO treatments have also been used for further purification of the BC microfibrils. For stage 3, once the pellicle is formed and purified, a sample is cut from the gel-like sheet. Water removal either by evaporation or a combination of pressing and evaporation collapses the gel-network and produces a dense film. For stage 4, chemical modification to the BC microfibril network can be achieved at three points along the engineered BC material processing, (i) during stage 1, (ii) during stage 2, and (iii) after stage 3 (i.e. to dried BC structures or films) [42].

Application

Cellulose is extensively used as a raw material in the paper industry in the production of paper and cardboard products. However, cellulose has shown its versatility in numerous applications.

Natural cellulose spheres are often applied in bioseparation, immobilized reaction, cell suspension culture, and as an adsorbent for sewage treatment. Spherical BC produced from dynamic method is translucent, loose, porous, and has a hydrophilic network structure. Its specific surface area increases with decreasing spherical diameter, so it could be used as a carrier to adsorb or crosslink various kinds of substances (e.g., enzyme, cell, protein, nucleic acid, and other compounds). Spherical BC may be applied in bioseparation, immobilized reaction, cell suspension culture, and as an adsorbent for sewage treatment. Compared with natural spherical cellulose, the fermentation production of BC spheres is simple, controllable and environment friendly. Moreover, BC sphere can be used repeatedly, expanding their potential applications.

Cellulosic derivatives such as cellulose acetate, cellulose propionate and cellulose acetate-butyrate, cast as membranes, have been reported as useful supports for immobilizing various enzymes such as catalase, alcohol oxidase and glucose oxidase. These supports gave better activity and storage stability for the enzymes. Cellulose ethers are widely used as important excipients for designing matrix tablets. On contact with water, the cellulose ethers start to swell and the hydrogel layer starts to grow around the dry core of the tablet. The hydrogel presents a diffusional barrier for water molecules penetrating into the polymer matrix and the drug molecules being released. Cellulose acetate butyrate microcapsules, as well as cellulose-based microspheres, have been used for the delivery of drugs [46].

Microbial cellulose synthesized by *Acetobacter xylinum* shows considerable potential as a novel wound healing system, resulting from its unique nanostructure. During the process of biosynthesis, various carbon compounds of the nutrition medium are utilized by the bacteria, then polymerized into single, linear -1,4-glucan chains and finally secreted outside the cells through a linear row of pores located on their outer membrane. Cellulose derived from *Acetobacter xylinum*, as discussed above in the context of wound healing, has also been explored as a potential scaffold material, due to its unusual material properties and degradability. Moreover, bacterial cellulose derived from *Acetobacter xylinum* has an ultrafine network architecture, high hydrophilicity, and mouldability during formation. In addition to the applications discussed, it is also suitable for use in micronerve surgery and as an artificial blood vessel suitable for microsurgery [47].

Chitin-Chitosan

Structure

Chitin is a white, hard, inelastic, nitrogenous polysaccharide found in the exoskeleton as well as in the internal structure of invertebrates.

Chitin is a hydrophobic linear polysaccharide derived from many natural sources including the exoskeleton of arthropods and insects and is the second most abundant natural polysaccharide next to cellulose. Chitin comprises a polysaccharide consisting of (1→4)- -N-acetyl-D-glucosamine units. Derivatives of chitin may be classified into two categories; in each case, the N-acetyl groups are removed, and the exposed amino function then reacts either with acyl chlorides or anhydrides to give the group NHCOR or is modified by reductive amination to $NHCH_2COOH$ of greatest potential importance are derivatives of both types formed by reaction with bi or polyfunctional reagents, thus carrying sites for further chemical reaction [48]. In practice, such reactions are carried out on native chitin or on incompletely deacetylated chitin, chitosan, so that the resulting polymer contains three types of monomeric units.

Chitosan is a partially deacetylated derivative of chitin and is the second most abundant biosynthesized material. Structurally, chitosan is a mixture of N-acetyl-D-glucosamine and D-glucosamine [49]. Generally, chitosan is insoluble in neutral or basic conditions, while protonation of free amino groups facilitates solubility of chitosan in dilute acids ($pH < 6$). In vivo degradation of chitosan is mainly attributed to the effect of lysozyme through hydrolysis of acetylated residues.

Chitosan itself chelates metal ions, especially those of transition metals, and also finds application as a matrix for immobilization of enzymes. Special attention has been given to the chemical modification of chitin, since it has the greatest potential to be fully exploited. Reactions with pure chitin have been carried out mostly in the solid state owing to the lack of solubility in ordinary solvents. A 50% deacetylated chitin has been found to be soluble in water [50]. This water soluble form of chitin is a useful starting material for its smooth modifications, through various reactions in solution phase. Some of the very recently reported chitosan derivatives are enumerated as follows:

- Fully deacetylated chitosan was treated with phthalic anhydride to give N-phthaloyl-chitosan. It was readily soluble in polar organic solvents. Further reactions had been carried out using this new derivative to improve the solubility of chitosan [51].
- To improve water solubility, Sashiwa *et al.* has successfully synthesized dendronized chitosan-sialic acid hybrids by using gallic acid as focal point and tri(ethylene glycol) as spacer arm. The

water solubility of these novel derivatives was further improved by N-succinylation of the remaining amine functionality [52].

- Recently, Baba *et al.* have synthesized methylthiocarbamoyl and phenylthiocarbamoyl chitosan derivatives to examine the selectivity toward metal ions from aqueous ammonium nitrate solution [53].
- The synthesis of chitosan hydrogels was carried out by Qu *et al.* by direct grafting of D,L-lactic and/or glycolic acid onto chitosan in the absence of catalysts. They demonstrated that a stronger interaction existed between water and chitosan chains after grafting lactic and/or glycolic acid. The side chains could aggregate and form physical crosslinking, which results in pH sensitive chitosan hydrogels [54].

Preparation

Chitin is easily obtained from crab or shrimp shells and fungal. In the first case, chitin production is associated with food industries such as shrimp canning. In the second case, the production of chitosan–glucan complexes is associated with fermentation processes, similar to those for the production of citric acid from *Aspergillus niger, Mucor rouxii,* and *Streptomyces,* which involves alkali treatment yielding chitosan–glucan complexes. The alkali removes the protein and deacetylates chitin simultaneously. Depending on the alkali concentration, some soluble glycans are removed. The processing of crustacean shells mainly involves the removal of proteins and the dissolution of calcium carbonate which is present in crab shells in high concentrations. The resulting chitin is deacetylated in 40% sodium hydroxide at 120^{0}C for 1–3 h. This treatment produces 70% deacetylated chitosan [55].

Application

Chitin and chitosan are known for their excellent biological properties, among which the biocompatibility with human cells, the ordered regeneration of wounded tissues, the immunoenhancing activity, the induction of immediate hemostasis, the radical scavenging activity, and the antimicrobial activity. Recent studies indicate that chitin and chitosan are most versatile in drug and gene delivery, elaborated diagnostics, devices for selective recognition of tumor cells, and surgical aids ranging from anti-adhesion gels to coated sterile stents.

As a kind of renewable resource, unmodified chitosan has been widely used in many fields such as pharmaceutical, agriculture, food, and biomedical applications. In order to realize the full potential of chitosan and bring a breakthrough in its broader utilization, attempts have been made to modify chitosan to obtain various derivatives. For the tissue repair and regeneration applications, chitosan can be functionalized by chemical reaction, coupling with specific ligands or moieties, combining with biomacromolecules, and crosslinking in the presence or absence of crosslinkers.

Particularly, skin substitute made of chitosan or its derivatives have attracted much attention due to the outstanding characteristics of chitosan, such as biocompatibility, hemostatic activity, antibacterial property, and ability to accelerate the wound-healing process [56].

The design of artificial kidney systems has made possible repetitive hemodialysis and the sustaining life of chronic kidney failure patients. Chitosan membranes have been proposed as an artificial kidney membrane because of their suitable permeability and high tensile strength. The most important part of artificial kidney is the semipermeable membrane and so far made from commercial regenerated cellulose and cuprophane. Since the primary action of the cellulose membrane is that of a sieve, there is little selectivity in the separation of two closely related molecules. These novel membranes need to be developed for better control of transport, ease of formability and inherent blood compatibility.

A series of membranes prepared from chitin and its derivatives improved dialysis properties. One of the most serious problems of using these artificial membranes is surface induced thrombosis, where heparization of blood is needed to prevent clotting, and people who are liable to internal hemorrhage can be dialysed only at great risk. Hence, these are the most challenging problem still to be resolved in the development of membranes which are inherently blood compatible. From these point of views, chitosan is hemostatic, i.e., causes clots [57].

Chitosan has replaced the synthetic polymers in opthalmological applications. Chitosan possesses all the characteristics required for an ideal contact lens; optical clarity, mechanical stability, sufficient optical correction, gas permeability, partially towards oxygen, wettability, and immunologically compatibility. Contact lenses are made from partially depolymerized and purified squid pen chitosan by spin casting technology, and these contact lenses are clear, tough, and possess other required physical properties such as modulus, tensile strength, tear strength, elongation, water content, and oxygen permeability. Antimicrobial and wound healing properties of chitosan along with excellent film forming capability make chitosan suitable for development of ocular bandage lens [58].

The special attention on chitosan has been paid for the repair of articular cartilage. Articular cartilage is particularly vulnerable to injury trama, disease or congenital abnormalities because of its avascular, alypmhatic and aneural nature. Once damaged, it has little capacity for intrinsic repair. Although many repair techniques have been attempted over the past four decades, but none has succeeded to regenerate long-lasting hyaline cartilage tissue to replace defected or damaged cartilage. Recently, preliminary studies on chitosan-GAG composite and its biologically interaction with articular chondrocytes showed promising results. Chitosan and its derivatives are being extensively used for bone tissue engineering and central nervous system also.

The growth of *Escherichia coli* was inhibited in the presence of chitosan. Chitosan also inhibited the growth of *Fusarium, Alternaria* and *Helminthosporium*. The cationic amino groups of chitosan probably bind to anionic groups of these microorganisms, resulting in growth inhibition. Extracellular lysozyme activity was enhanced in in vitro cultures of several mammalian cells by treatment with chitin and its derivatives. As a result, connective tissue formation was stimulated, and the self-defence function against microbial infection was enhanced at the cellular level. On the basis of these results, several chitin and chitosan dressing materials have been developed commercially for the healing treatment of human and animal wounds [59].

Chitosan is non-toxic and easily bioabsorbable with gel-forming ability at low pH. Moreover, chitosan has antacid and antiulcer activities which prevent or weaken drug irritation in the stomach. Also, chitosan matrix formulations appear to float and gradually swell in an acid medium. All these interesting properties of chitosan make this natural polymer an ideal candidate for controlled drug release formulations [60].

DECELLULARIZATION

Autologous grafts are "gold standard" for implantation. However, the most disadvantage of autologous is quantity. The number of autograft does not meet needs of patients. Homograft is greater than autograft but they cannot satisfy needs of patient. Many patients must wait for a long time to take a homogenous organ. Xenograft is the greatest but they can evoke serious immune reaction. So, one method developed to process homograft and xenograft is decellularization. Every tissue/organ concludes cells and extracellular matrix. Cells are structure and functional units of tissue/organ but cells are major antigen of tissue/organ. Extracellular matrix is many protein, polysaccharide, protoglycan released by cell. ECM plays an important role in mechanical support, signal transportation, adherence of tissue/organ. Decellularization is a multi-step process to remove all cell components from tissue/organ and leave intact ECM. Many decellularization agents were researched such as physical methods, chemical methods and enzyme methods. Every decellularization agent has specific affections of cell and extracellular matrix. So, these agents are combined to make an effective decellularization process which removes all cell components and reverses maximum ECM. Decellularization effectiveness depends on type of tissue/organ. One agent can be a good detergent for decellularizing one tissue but not for another [61]. Moreover, cell derived ECM can be used as a matrix for cell culture.

Table 1: Overview of decellularization methods (Thomas W. Gilbert)

Method	Mode of action	Effects on ECM
Physical		
Snap freezing	Intracellular ice crystals disrupt cell membrane	ECM can be disrupted or fracture during rapid freezing
Mechanical force	Pressure can burst cells and tissue removal eliminates cells	Mechanical force can cause damage to ECM

Mechanical agitation	Can cause cell lysis, but more commonly used to facilitate chemical exposure and cellular material removal	Aggressive agitation or sonication can disrupt ECM as the cellular material is removed
Chemical		
Alkaline; acid	Solubilize cytoplasmic components of cells; disrupts nucleic acids	Removes GAGs
Non-ionic detergents		
Triton X-100	Disrupts lipid–lipid and lipid–protein interactions, while leaving protein – protein interactions intact	Mixed results; efficiency dependent on tissue, removes GAGs
Ionic detergents		
Sodium dodecyl sulfate (SDS)	Solubilize cytoplasmic and nuclear cellular membranes; tend to denature proteins	Removes nuclear remnants and cytoplasmic proteins; tends to disrupt native tissue structure, remove GAGs and damage collagen
Sodium deoxycholate		More disruptive to tissue structure than SDS
Triton X-200		Yielded efficient cell removal when used with zwitterionic detergents
Zwitterionic detergents		
CHAPS	Exhibit properties of non-ionic and ionic detergents	Efficient cell removal with ECM disruption similar to that of Triton X-100
Sulfobetaine-10 and -16 (SB-10, SB-16)		Yielded cell removal and mild ECM disruption with Triton X-200

Tri(n-butyl) phosphate	Organic solvent that disrupts protein–protein interactions	Variable cell removal; loss of collagen content, although effect on mechanical properties was minimal
Hypotonic and hypertonic solutions	Cell lysis by osmotic shock	Efficient for cell lysis, but does not effectively remove the cellular remnants
EDTA, EGTA	Chelating agents that bind divalent metallic ions, thereby disrupting cell adhesion to ECM	No isolated exposure, typically used with enzymatic methods (e.g., trypsin)
Enzymatic		
Trypsin	Cleaves peptide bonds on the C-side of Arg and Lys	Prolonged exposure can disrupt ECM structure, removes laminin, fibronectin, elastin, and GAGs
Endonucleases	Catalyze the hydrolysis of the interior bonds of ribonucleotide and deoxyribonucleotide chains	Difficult to remove from the tissue and could invoke an immune response
Exonucleases	Catalyze the hydrolysis of the terminal bonds of ribonucleotide and deoxyribonucleotide chains	

Today, many decellularization grafts were applied in clinical treatment such as acellular valve, vascular and some are commercial include: SurgisSIS (porcine small intestinal submicosa), Alloderm (human dermis), ACell (porcine urinary bladder)... Small intestine contains four layers such as mucosa, submucosa, muscularis externa and serosa. Small intestine submucosa (SIS) is a submucosa tissue between mucosa and muscularis externa. SIS is isolated from small intestine by mechanically removing of internal mucosal layer and outer muscular layer. Then, SIS is processed step-by-step with 0.1% peracid acid, 0.05% gentamycin and sterilized using 2500 kRad gamma irradiation. SIS is consisted of collagen, proteoglycan, glycosaminoglycan, glycoprotein and growth factor (VEGF, FGF-2 ...). Most of these components can be preserved in extracellular matrix after decellularizing. In tissue engineering, SIS is used as soft tissue grafts such as vascular, skin or used for reconstructing genitourinary, ligament tissue [62].

Vascular Tissue Engineering

Atherosclerosis is the most dangerous cardiovascular disease. Atherosclerosis is a condition in which vascular wall is harden, thicken because of fatty lipid accumulation. Atherosclerosis is responsible for millions of death all the world every year. In 2004, seventeen million people passed away because of cardiovascular disease all over the world (29% world total death). More than 7 million people are killed by coronary disease (9.6% world total death) (WHO). Heart American Association, in 2006, coronary disease was responsible for 17.6% death in America. One out of six American died for coronary disease [63].

Today, there are some methods to cure atherosclerosis including drug treatment, coronary artery bypass surgery and angioplasty. However, these methods can not cure this condition completely. More than 30% patients don't have qualified autogenous vascular grafts. Therefore, artificial grafts are made to overcome some current drawbacks.

Artifical graft can be classified into synthetic and biologic scaffold. Synthetic scaffolds include undegradable polymers such as Dacron, ePTFE and degradable polymers such as polylactic acid (PLA) and polyglycolic acid (PGA). Some advantages of synthetic polymers are easy processing, high initial strength, however, they have some disadvantages such as slow recellularization, poor ability to support remodeling. Biologic vascular grafts are either obtained by manipulating native proteins of vascular vessels such as collagen, elastin... or by decellularizing vascular vessels. Decellularized native vascular vessels have some advantages such as slow cost, reduction of graft rejection and immune reaction [64].

In the early time, decellularization studies were tested without cell-seeding. In 1990, Lantz et al use SIS (small intestine submucosa) as small-diameter arterial graft in18 dogs. 48 weeks after surgery,

28 vascular grafts (75%) were patent and tree dogs can survive from 76 to 82 weeks after implantation. This result suggested that small intestinal submucosa can be used in small diameter vascular tissue engineering [65]. In 2001, Hodde et al demonstrated that porcine SIS ECM extract containing vascular endothelial growth factor (VEGF) with concentration of 0.77 ng VEGF/g SIS [66]. In the same year, Nemcova et al isolated and decellularized SIS from bovine and porcine. Nine decellularized SIS grafts were implanted into femoral arteries of five male mongrel dogs. After 9 weeks, eight grafts remained patent and some kinds of cell such as endothelial cells (ECs), smooth muscle cells (SMCs) invaded into the grafts. No evidence of inflammatory and aneurysmal symptom was observed during the experimental time [67].

However, the main disadvantage of vascular graft is thrombus which ussually occurs immediately in vascular graft lumen after implantation. This process can lead to graft failure and threaten patients' survival. In order to solve this problem, ECs are used as anti-thrombus agent in vascular graft lumen, additionally, SMCs are used to improve mechanical strength of vascular graft. Broschel et al decellularized and recellularized rat iliac arterial grafts. Iliac arterial grafts were decellualrized by glycerin, SDS for 12 hours. Then, decellularized iliac grafts were recellularized with adult rat heart ECs and implanted to femoral arteries of allogeneic rats without systemic anticoagulation injection. After 4 weeks, 2 of 7 control grafts (29%) were patent (decellularized grafts without recellularization) and 8 of 9 (89%) experimental grafts (decellularized grafts with recellularization) maintained blood vessel patent [68]. Consequently, this experiment result proved excellent function of ECs in vascular tissue engineering. Some researchers on vascular tissue engineering seeded autologous vascular cells including ECs, SMCs and fibroblasts to make autologous tissue engineered vascular. At the same time, the appearance of bioreactor systems sped up development of vascular tissue engineering. Niklason developed a pulsatile perfusion bioreactor system in 2001. Bioreactor contained a stirbar, a lid for gas exchange and one (or two) silicone tubing(s). Porous vascular scaffolds were threaded over silicone tubing. Bovine aortic SMCs at the passage 2 or 3 were pipetted onto the outer surface of the scaffolds, then bioreactors were slowly rotated and removed to incubator with 10% CO_2, 100% humidity, and 37°C. Each silicone tubing was linked to a pulsatile perfusion system operated at 165 beats/minute and 260/-30 mmHg. After 8 weeks culture, under nonpulsatile condition, SMC growth was on the outer surface of tubing scaffold and poorly organized. Under pulsatile condition, SMC distribute homogenously in scaffold wall from outer to inner similar to native structure [69].

Nowaday, some decellularized vascular grafts can be obtained from human. Human umbilical veins and arteries can be used in decellularization experiment. Human umbilical arteries were isolated and completely decellularized by Gui et al in 2009. Decellularized umbilical arteries preserved intact collagen matrix and mechanical properties, burst pressure had no signicant change from native form. Decellularized artery graft remained patent after 8 weeks surgery [70]. In 2005, Joel Daniel et al processed human umbilical vein by automated dissection. Human umbilical cord was inserted a mandrel into vein and frozen to -20 and -80°C, human umbilical cord was maintained for 12 hours at least at this temperature. The mandrel was lathed with rotation speed of 2900 rpm, cutting depth of 750 μm. Then, human umbilical vein was decellularized with 1% (w/v) SDS. The result showed that the treated human umbilical vein contained no cell; burst pressure results were 972.8 ± 133.8 mmHg (972/1082), compliance results were 5.7 ± 1.3% over 80 – 120 mmHg. Vascular smooth muscle cells can adhere, proliferate and migrate on the surface of dHUV [71]. Tran Le Bao Ha et al carried out a research on decellularization method for HUVs. HUVs were isolated by manual dissection. HUVs were decellularized by distilled water or NaCl 3M or SDS 1%. The result confirmed that the combination between SDS 1% (24h) and NaCl 3M (24h) showed the best effective on cellular elimination.

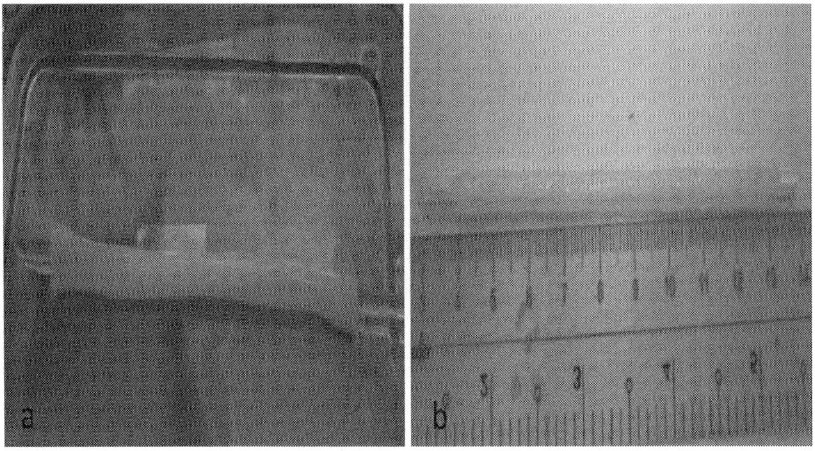

Figure 6. Human umbilical cord (a) and human umbilical vien (b).

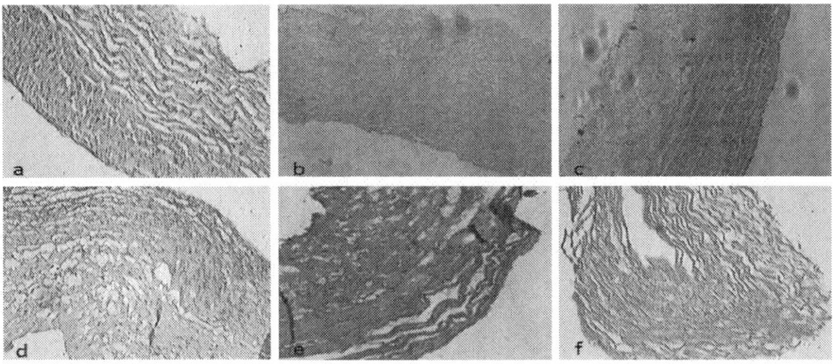

Figure 7: HE staining of HUV with different decellularization methods. Control (a), distilled water (b), NaCl 3M (c), SDS 1% 24h (d), SDS 1% 36h (e), SDS 1% 24. and NaCl 3M 24h (f).

ECM From Cultured Cells

One of the most important properties of ECM is its functional diversity. ECM has been reported to support and enhance for adhesion, migration, proliferation of cells as well as to create stem cell niches*in vitro*.. ECM can be harvested from different sources, one of which is from cells under culture condition. When cultured, cells will produce three-dimensional matrix surrounding themselves. A method is described for generating tissue culture surfaces coated with a human fibroblast-derived ECM [72, 73, 74, 75].

For this purpose, human foreskin fibroblasts are isolated, plated and cultured until third passage (seeFigure 8). Fibroblasts are maintained in culture medium until reaching 80% confluency and stimulated to synthesize ECM by culture medium supplemented ascorbic acid. Matrices are denuded of cells and cellular remnants are removed by using Triton X-100, NH_4OH and DNase.

Then, ECM coated culture surfaces are tested by staining with PI to access DNA remnant (see Figure 9), with H&E and PAS in order to characterize component of ECM. The results suggest that fibroblast-derived three-dimensional matrix was determined to be free of cellular constituents and still remain attached to the culture surface. The conducted matrices were washed and covered with PBS; and stored at 4^0C. Under these conditions, biological activity (for example, induction of cell attachment, proliferation...) was reported to be well-preserved for up to 6 months.

In order to evaluate the quality of the ECM, the test of cell rapid attachment ability is performed. Cell proliferation on ECM is also assessed. Results showed that the cultured cells attached and proliferated on ECM coated surfaces faster than on ECM non-coated surfaces.

Figure 8. Fibroblasts are in cultured surfaces (a) and are stimulated to synthesize ECM (b).

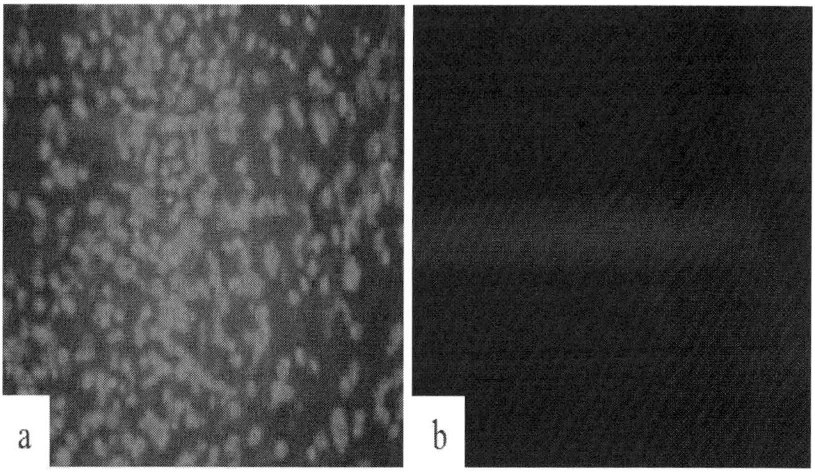

Figure 9.ECM stained PI before (a) and after (b) using DNase.

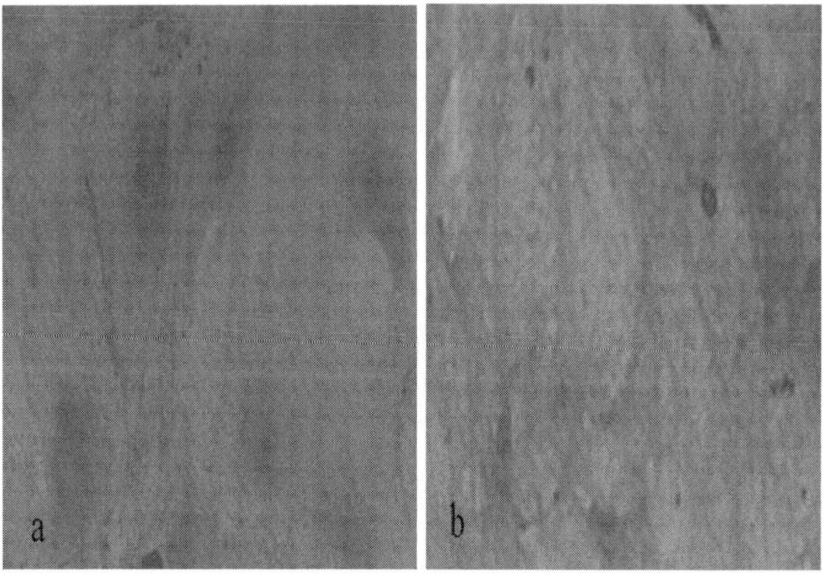

Figure 10. Culture surface before (a) and after (b) is denuded cells.

CONCLUSION

In modern society, many diseases have been increasing in human because of pollution, accident, lifestyle... The mutilation in human body leads to expand the needs of replacing tissues/organs. However, the available source of tissues/organs is limited. Creating artificial tissues/organs for replacing damaged, dysfunctional tissues/organs becomes a big discipline on material science. Naturally derived biomaterials have been studied and applied in clinical applications as artificial tissues/organs because they are capable of supporting cell proliferation, biodegradability and remodeling tissues. Although the current results have not completely satisfy the clinical demand, the potential applications of naturally derived biomaterials are still highly considered, therefore, research on this field have now being taken place all over the world.

REFERENCES

1. H Lodish, A Berk, S. L Zipursky, Molecular Cell Biology. New York: W. H. Freeman 2000
2. M Anu, Type XV collagen. University of Oulu 2001
3. N Hassan, P Habibollah, A Abolhassan, Properties of the amniotic membrane for potential use in tissue engineering. European cells & materials 2008158889
4. H Andrew, A. S Vijay, G Trevor, M. Y Aaron, L James, K. J David, S. D Harminder, Optimization of amniotic membrane (AM) denuding for tissue engineering. Tissue engineering 200814371383
5. R Singh, S Purohit, M. P Chacharkar, Effect of high doses of gamma radiation on the functional characteristics of amniotic membrane. Radiation Physics and Chemistry 20077610261030
6. M Loss, V Wedler, W Kunzi, C Meuli-simmen, V. E Meyer, Artificial skin, split-thickness autograft and cultured autologous keratinocytes combined to treat a severe burn injury of 93% of TBSA. Burns 200026644652
7. K. B Ludwik, N. H David, M. C Mario, B. N William, E. M Oscar, G. J Marc, Amnion in the treatment of pediatric partial-thickness facial burns. Burns 200834393399
8. S. Y Susan, Current Protocols in Cell Biology. John Wiley & Sons, Inc. 200312
9. P. J Rose, H. F Mark, N. M Bikales, C. G Overberger, G Menges, J. I Kroschwitz, editors. Encyclopedia of polymer science and engineering. New York, USA: Wiley 1989488513
10. K. I Draget, O Smidsrod, G Skjàk-bræk, Alginate from algae. In Polysaccharide and Polyamides in the food Industry. In: Steinbüchel A, Rhee SK. (eds). Wiley-VCH Verlag GmbH & Co.: Weinheim 2005130
11. H Ertesvåg, S Valla, G Skjåk-bræk, Genetics and biosynthesis of alginates. Carbohyd. Europe 1996141418
12. W Paul, C. P Sharma, Chitosan and Alginate Wound Dressings: A Short Review. Trends in Biomaterials and Artificial Organs 2004181823

13. J.PChen,I.MChu,M.YShiao,B.RHsu,S.HFu,Microencapsulation of islets in PEG amine modified alginate-poly(l-lysine)-alginate microcapsules for constructing bioartificial pancreas. Journal of Fermentation and Bioengineering 199886185190
14. J. A Rowley, G Madlambayan, D. J Mooney, Alginate hydrogels as synthetic extracellular matrix materials. Biomaterials 1999204553
15. D. B Seifert, J. A Phillips, Porous alginate-poly(ethylene glycol) entrapment system for the cultivation of mammalian cells. Biotechnology progress 199713569576
16. G.HAltman,Silk-basedbiomaterials.Biomaterials200324401416
17. K Tanaka, S Inoue, S Mizuno, Hydrophobic interaction of 25containing Asn-linked oligosaccharide chains, with the H-L complex of silk fibroin produced by B. mori. Insect Biochemistry and Molecular Biology 1999
18. S Inoue, Silk fibroin of Bombyx mori is secreted, assembling a high molecular mass elementary unit consisting of H-chain, L-chain, and 25with a 6:6:1 molar ratio. Journal of Biological Chemistry 2000
19. A Seidel, Regenerated spider silk: Processing, properties, and structure. Macromolecules 200033775780
20. M Xu, Structure of a protein superfiber: spider dragline silk. Proceedings of the National Academy of Sciences of the United States of America 19908771207124
21. P Kumaresan, Sericin a versatile by product. Indian Silk 2007451113
22. S. B Dandin, Biomedical uses of silk and its derivatives. Indian Silk 20074558
23. M. L Gulrajani, Sericin-a bio-molecule of value. Indian Silk 2006451622
24. Y Wang, Cartilage tissue engineering with silk scaffolds and human articular chondrocytes. Biomaterials 20062744344442
25. S Sofia, Functionalized silk-based biomaterials for bone formation. Journal of Biomedical Materials Research 200154139148
26. Z Megeed, Genetically engineered silk-elastinlike protein polymers for controlled drug delivery. Advanced drug delivery reviews 20025410751091

27. L. V Medved, O. V Gorkun, P. L Privalov, Structural organization of C-terminal parts of fibrinogen A alpha-chains. FEBS Lett 1983160291295
28. J. D Ferry, P. R Morrison, Preparation and properties of serum and plasma proteins. VIII. The conversion of human fibrinogen to fibrin under various conditions. Journal of the American Chemical Society 194769388400
29. L Betts, B. K Merenbloom, S. T Lord, The structure of fibrinogen fragment D with the 'A' knob peptide GPRVVE. Journal of Thrombosis and Haemostasis 2006411391141
30. E. A Ryan, L. F Mockros, A. M Stern, L Lorand, Influence of a natural and a synthetic inhibitor of factor XIIIa on fibrin clot rheology. Biophysical Journal 19997728272836
31. J. P Collet, H Shuman, R. E Ledger, S Lee, J. W Weisel, The elasticity of an individual fibrin fiber in a clot. Proceedings of the National Academy of Sciences of the United States of America 200510291339137
32. C Kaitlin, J Murphy, Kent Leach. A reproducible, high throughput method for fabricating fibrin gels. BMC Research Notes 2012
33. W Bensaïd, J. T Triffitt, C Blanchat, K Oudina, L Sedel, H Petite, A biodegradable fibrin scaffold for mesenchymal stem cell transplantation. Biomaterials 20032424972502
34. M Okada, B Blomback, Calcium and fibrin gel structure. Thrombosis Research 198329269280
35. K. R Siebenlist, M. W Mosesson, Progressive cross-linking of fibrin gamma chains increases resistance to fibrinolysis. Journal of Biological Chemistry 19942692841428419
36. R. H Alan, K. G Dennis, P. H Mark, C. S Frank, E Constantine, Autologous Whole Plasma Fibrin Gel: Intraoperative Procurement. Archieves of Surgery 1992127357359
37. Q Ye, G Zünd, P Benedikt, Fibrin gel as a three dimensional matrix in cardiovascular tissue engineering. European Journal of Cardio-thoracic Surgery 200017587591
38. W Bensa, J. T Triffittb, C Blanchata, K Oudinaa, L Sedela, H Petitea, A biodegradable fibrin scaffold for mesenchymal stemcell transplantation. Biomaterials 200324972502

39. R Saltz, A Dimick, C Harris, J. C Grotting, J Psillakis, L. O Vasconez, Application of autologous fibrin glue in burn wounds. Journal of Burn Care & Rehabilitation 198910504507
40. D Klemm, B Philipp, T Heinze, U Heinze, W Wagenknecht, Comprehensive Cellulose Chemistry. Fundamentals and Analytical Methods. Wiley-VCH, Weinheim 1998
41. V. V Myasoedova, Physical chemistry of non-aqueous solutions of cellulose and its derivatives. John Wiley and Sons, Chirchester 2000
42. J. M Robert, M Ashlie, N John, S John, Y Je, Cellulose nanomaterials review: structure, properties and nanocomposites. Chemical Society Reviews 20114039413994
43. G. M Nabar, C V Padmanabhan, Studies in oxycellulose. Proceedings of the Indian Academy of Sciences 195031371380
44. A. H Kibbe, Handbook of pharmaceutical excipients: Cellulose, silicified microcrystalline. American Public Health Association. Washington 2000
45. L. C Wadworth, D Daponte, Cellulose esters. In: Nevell TP, Zeronian SH (ed.) Cellulose Chemistry and its Applications. Ellis Horwood, Chich-ester 1985349362
46. J. N Delgado, A William, Wilson and Gisvold's textbook of organic medicinal and pharmaceutical chemistry. Lippincott-Raven Publishers, Wickford 1998
47. A Svensson, E Nicklasson, T Harrah, B Panilaitis, D Kaplan, M Brittberg, P Gatenholm, Bacterial Cellulose as a Potential Scaffold for Tissue Engineering of Cartilage. Biomaterials 200526419431
48. Muzzarellieditor. Natural Chelating Polymers. Pergamon Press, New York 1973
49. Zikakiseditor. Chitin, Chitosan and Related Enzymes. Academic Press, Orlando 1984
50. P. K Dutta, Ravikumar MNV, Dutta J. Chitin and chitosan for versatile applications, Journal of Macromolecular Science-polymer Reviews 200242307315
51. K Kurita, Chemistry and application of chitin and chitosan, Polymer Degradation and Stability 199859117120
52. H Sashiwa, Y Shigemasa, R Roy, Chemical modification of chitosan: synthesis of dendronized chitosan-sialic acid hybrid

using convergent grafting of preassembled dendrons built on gallic acid and tri(ethylene glycol) backbone. Macromolocules 20013439053920

53. Y Baba, H Noma, R Nakayama, Y Matsushita, Preparation of chitosan derivatives containing methylthiocarbamoyl and phenylthiocarbamoyl groups and their selective adsorption of copper (II) over iron (III). Analytical sciences 200218359370
54. X Qu, A Wirsen, A. C Albertsson, Effect of lactic/glycolic acid side chains on the thermal degradation kinetics of chitosan derivatives. Polymer 20014148414850
55. Madhavan editor. Chitin Chitosan and their Novel Applications. Science Lecture Series, CIFT, Kochi 1992
56. M Prasitslip, R Jenwithisuk, K Kongsuwan, N Damrongchai, P Watts, Cellular responses to chitosan in vitro: the importance of deacetylation, Journal of Materials Science: Materials in Medicine 200011773780
57. W. C Lin, T. Y Liu, M. C Yang, Hemocompatibility of polyacrylonitrile dialysis membrane immobilized with chitosan and heparin conjugate. Biomaterials 2003
58. Markey Bowman, Bergamini, editor. Chitin and Chitosan. Elsevier Applied Science, London 1989
59. Gebelein CarraherJr, editor. Industrial Biotechnological Polymers. Technomic, Lancaster 1995
60. K. E Uhrich, S. M Cannizzaro, R. L Langer, K. M Shakesheff, Polymeric systems for controlled drug release. Chemical Reviews 19999931813190
61. W. G Thomas, Decellularization of tissues and organs, Biomaterials 20062736753683
62. H Jason, Naturally Occurring Scaffold for Soft Tissue Repair and Regeneration. Tissue Engineering 20028295308
63. L. R Véronique, Heart Disease and Stroke Statistics- 2012Update: A Report from the Amercian Heart Association. Circulation 2012.
64. W Chrysanthi, Engineering of Small Diamter Vessels.Therapeutic applications: tissue therapy; 10001013
65. Lantz Small Intestinal Submucosa as a Small-Diameter Arterial Graft in the Dog. Journal of Investigative Surgery 19903217227

66. J. P Hodde, Vascular endothelial growth factor in porcine-derived extracellular matrix. Endothelium 200181124
67. S Nemcova, A. A Noel, C. J Jost, P Gloviczki, V. M Miller, K. G Brockbank, Evaluation of a xenogeneic acellular collagen matrix as a small-diameter vascular graft in dogs preliminary observations. Journal of Investigative Surgery 2001
68. G. H Borschel, Tissue engineering of recellularized small-diameter vascular grafts. Tissue Engineering 200511778786
69. E. N Laura, Morphologic and mechanical characteristics of engineered bovine arteries. Journal of vascular surgery 200133628638
70. L Gui, Development of decellularized human umbilical arteries as small-diameter vascular grafts. Tissue Engineering 20091526652676
71. D Joel, Development of the Human Umbilical Vein Scaffold for Cardiovascular Tissue Engineering Applications. ASAIO Journal 200551252261
72. M. Y Kenneth, Current Protocols in Cell Biology. John Wiley & Sons, Inc. 2003110
73. S Badylak, T Gilbert, J Irvin, The extracellular matrix as a biologic scaffold for tissue engineering. Tissue Engineering 2008121144
74. R. N Chen, H. O Ho, Y. T Tsai, M. T Sheu, Process development of an acellular dermal matrix (ADM) for biomedical application. Biomaterials 20042526792686
75. Z. J Cheng, R. P So, H. C Byung, P Kwideok, H. M Byoung, *In vivo* cartilage tissue engineering using a cell-derived extracellular matrix scaffold. Artificial Organs 200731183192

Citations

CHAPTER 1

Awajiogak Anthony Ujile and Dirina Amesi, "Performance Evaluation of Refrigeration Units in Natural Gas Liquid Extraction Plant," Journal of Thermodynamics, vol. 2014, Article ID 863408, 7 pages, 2014. doi:10.1155/2014/863408.

CHAPTER 2

Mohammad O. Eshkalak, Shahab D. Mohaghegh, and Soodabeh Esmaili, "Geomechanical Properties of Unconventional Shale Reservoirs," Journal of Petroleum Engineering, vol. 2014, Article ID 961641, 10 pages, 2014. doi:10.1155/2014/961641.

CHAPTER 3

Vyacheslav V. Kalashnikov, Gerardo A. Pérez-Valdés, Timothy I. Matis, and Nataliya I. Kalashnykova, "US Natural Gas Market Classification Using Pooled Regression," Mathematical Problems in Engineering, vol. 2014, Article ID 695084, 9 pages, 2014. doi:10.1155/2014/695084.

CHAPTER 4

Mohan, Jiang, W Michael Griffin, Chris Hendrickson, Paulina Jaramillo, Jeanne VanBriesen, and Aranya Venkatesh, Life cycle greenhouse gas emissions of Marcellus shale gas, doi:10.1088/1748-9326/6/3/034014.

CHAPTER 5

Shonkoff SB, Hays J, Finkel ML. 2014. Environmental public health dimensions of shale and tight gas development. Environ Health Perspect 122:787–795;http://dx.doi.org/10.1289/ehp.1307866.

CHAPTER 6

Botao Qu, Xinrong Lu, Yan Wu, Xiaozeng You, and Xiangxing Xu, Synthesis of copper micro-rods with layered nano-structure by thermal decomposition of the coordination complex $Cu(BTA)_2$, doi:10.1186/s11671-015-0769-7.

CHAPTER 7

Reza Abedini and Amir Nezhadmoghadam, Application of Membrane in Gas Separation Processes: its Suitability and Mechanisms, ISSN 1337-7027.

CHAPTER 8

Muhammad Khan Memon, Saleem Qadir Tunio, Khalil Rehman Memon, and Arshad Ahmed Lashari, A Comparative Study of Liquefied Natural Gas: An Overview, ISSN: 2040-7459; e-ISSN: 2040-7467.

CHAPTER 9

Tran Le Bao Ha, To Minh Quan, Doan Nguyen Vu and Do Minh Si (2013). Naturally Derived Biomaterials: Preparation and Application, Regenerative Medicine and Tissue Engineering, Prof. Jose A. Andrades (Ed.), ISBN: 978-953-51-1108-5, InTech, DOI: 10.5772/55668.

Index